The Physics of
HOLLYWOOD

Using current Hollywood movies to inspire teaching

Jan-Martin Klinge & Samuel Cardeña

3. Edition

The authors:

Jan-Martin Klinge was born in 1981 and has been working at a secondary school near Siegen, Germany, since 2012. He is the author of the teacher's blog 'Halbtagsblog.de', where he extensively writes about the developments of new teaching methods. Alongside his work, Klinge also writes for other types of media and shares his experiences of work at conferences. For his dedication he has been honored multiple times a Microsoft Innovative Education Expert (MIE Expert).

Samuel Cardeña Sánchez was born in 1984 and currently works in a secondary school in Ponce, Puerto Rico, since 2015. He has been teaching since 2008 and has been able to participate in multiple related projects in physics, mathematics and robotics. In addition, Cardeña is also an educational consultant in STEM areas for the Caribbean. His latest educational project, history of Puerto Rico through the lens of robotics, has allowed him to travel to different parts of the world through Microsoft as an innovative educational teacher (MIE expert).

Translated by Joana Czycholl

www.halbtagsblog.de

Dear Reader,

Of all school subjects Chemistry and Physics are known to be the least liked subjects by far. They are the first to be dropped and their contents are the quickest to be forgotten.

Today we are in the absurd situation of living in a highly technical world, surrounded by physics, without being able to find a real approach to the subject. Things are supposed to 'just work'.
The hurdles of understanding are already identifiable in school: In lessons, old equipment is being taken out of old cupboards in order to investigate strange phenomena which are given new, strange definitions/terms. A lot of physical concepts are initially abstract and have little to do with everyday life: 'What exactly is an electric field?', 'What is it that makes a magnetic field magnetic?', 'What is the difference between mass and weight?', 'What is a moment of force/torque?'.
This is not just hard to grasp for children.

What is more, in contrast to Mathematics in Physics we have to deliver far more abstract and complex content in far less time. While in Mathematics we have a few weeks to spend on percentage calculation, in Physics we have roughly the same amount of time to introduce the Newtonian axioms discussed, the lever explained, and the pulley experimentally explored. In a subsidiary subject and if unlucky this happens on Tuesday afternoon during last period. This usually leaves little time for projects or activities that are fun or motivating.

Incorporating movie sequences into the lesson links the student's world with the world of Physics. Whether Spiderman throws himself off a roof or a marble rolls off the table – the same physical principles apply. Momentum conservation can be found in Newton's pendulum as well as in Arnold Schwarzenegger's action films when he wildly shoots everything in sight. But I can imagine what the kids might be more interested in.

As far as possible on the following pages I would like to steer clear of pedagogical and didactical derivation.

As a teacher you will have studied and gathered experience long enough to know when the showing of a film sequence will or will not be sensible or useful.

If you are no teacher this will not be of interest to you anyway.

As a physicist you also will not need the derivation of the formula for a thread pendulum. Such things are better found within appropriate technical literature.

To non-physicists I recommend reading the books by Metin Tolan or 'The Physics of Superheros' by James Kakalios. In these the authors steer clear of using Mathematics (as far as possible). (In case you have already bought this book as a non-teacher or physicist I am very sorry – maybe gift it to a teacher within your circle of friends.)

This book is meant to focus on concrete lesson examples. In my imagination you are a teacher in secondary school or college looking, for practical ideas to convey topics within physics in an interesting way. The following examples are therefore reduced to the teaching essential. This is not about calculating the world but about linking educational physics with movie sequences in order to make it more accessible to young people.

For every task I have outlined a short approach: I am also sure for every task there is a better, more accurate approach and I invite you to dismiss my solution and find one that works best for your lesson. In the event that you would like to ponder a question yourself, the calculations are never on the same page as the Solution. Spoiler alert.

The examples are supposed to be exemplary. If you flip through them and keep the different assignments of tasks in mind you will keep noticing more and more movie excerpts that could be used in physics lessons. This book is meant to increase your awareness and give you ideas.

For lessons it is advisable to intersperse tasks here and there. I myself have also had good experiences with combining a number of different tasks into what I call 'The Learning Market' (more on that here: https://halbtagsblog.de/schule/lerntheken/english-learning-market/) with which it is easy to repeat the content at the end of the unit. The students chose individual tasks and calculated them with pleasure. I reported this in depth on my blog where you can also find further examples.

Now we hope you enjoy reading this book.

April 2019   Jan-Martin Klinge & Samuel Cardeña

SPIDER-MAN 3

Keywords: Mechanics, Freefall, Braking Effort, Velocity, Force
Time of movie excerpt: 01:44:30 – 01:48:40
Movie Trailer: https://www.youtube.com/watch?v=ukGsu63s2dk

Content

In the third Spiderman film published in 2007 the hero, played by Tobey Maguire, fights his nemesis 'Venom' in skyscraper canyons. Both cling to the wall of a building but when it collapses Spiderman falls about 80 meters before catching himself with a string of spider silk and continuing his fight with Venom. In the film it can be seen that the braking distance is about 20 meters.

Tasks

1) How much time (?) does the 80-meter fall take?

2) How fast does Spiderman fall in the end?

3) Calculate the braking effort Spiderman's arm is exposed to.

4) From the horrid medieval punishment of quartering we can estimate that an arm will rip off at a physical strain of about 3000 Newton. Can Spiderman withhold this strain?

SOLUTION

1. From $s = \frac{1}{2} \cdot a \cdot t^2$ it follows that $t = \sqrt{\frac{2 \cdot s}{a}} = \sqrt{\frac{160\ m}{9,81\ m/s^2}} = 4s$

2. It holds that $v = a \cdot t = 39.6 \frac{m}{s} = 142.6 \frac{km}{h}$

3. The braking effort has to balance the whole of the kinetic energy. The kinetic energy (estimating Spiderman's mass at 70kg) amounts to:

$$E_{kin} = \frac{1}{2} \cdot m \cdot v^2 = \frac{1}{2} \cdot 70kg \cdot \left(39.6 \frac{m}{s}\right)^2 = 54885\ J$$

The braking effort $W_{Brake} = F \cdot s$ is as large as the kinetic energy, solved to F it holds that:

$$F = \frac{W_{Brake}}{s} = \frac{E_{kin}}{s} = 2744\ N$$

Therefore, Spiderman's arm is exposed to a force of 2744 N which equals a weight of close to 300kg.

4. The arm would not be pulled off (just yet), but it would surely pop out of its socket which would definitely be very painful. In any case, Spiderman should probably use both of his arms.

SKULL ISLAND

Keywords: Volume, weight, mass, magnification/expansion
Time of movie excerpt: s. Trailer
Movie Trailer: https://www.youtube.com/watch?v=a3s5LhjaWoo

Content

King Kong was one of the first monsters created for film and not adapted from literature. The original was produced in 1933, in 2017 a spectacular remake was published. Just the trailer alone already poses a number of exciting questions.

Tasks

1) Approximately how tall is Kong in the trailer?

2) In reality gorillas stand at about 1.5m and can weigh up to 200kg. So to what factor have length, width and depth changed?

3) Give King Kong's weight. To what factor has it changed?

4) A body is supported by its skeleton. How much weight a bone can carry also depends on its cross section. A human thigh bone has a diameter of about 5cm and can withstand a strain of about 15000 N. To what factor does the cross section of King Kong's bones change in relation to a normal gorilla?

5) Compare the results from 3) and 4). What does this mean for King Kong in reality?

SOLUTION

1) The height of the ape can roughly be estimated in the scene in which he wades through the water: Kong will rise about 25 meters up into the air.

2) The factor is $\frac{25}{1,5} = 16.7$

3) The weight changes (analogous to volume) with the third power and therefore measures around 926 tons.

4) The cross section of the thigh grows with the second power only. Since human and gorilla are roughly similar, the cross section of Kong's thigh measures 16.7^2 times that of a normal gorilla. Analogous the thigh could withhold a strain of around $4 \cdot 10^6 \, N$ which equals about 400 tons.

5) Kong's thighs are not able to carry the weight of his body.

Addition: You could add another calculation here: *How big can King Kong be at maximum so that his thighs can still support his weight?*

The Physics of Hollywood

GOTHAM

Keywords: Force, Parallelogram of forces
Time of movie excerpt: s. Trailer
Movie Trailer: https://www.youtube.com/watch?v=SgMEyWN_BXY

Content

The television series Gotham (2016) shows the youth of the billionaire Bruce Wayne who later goes on the hunt for criminals as superhero 'Batman'. In one episode of series three, him and his accomplice Selina Kyle are forced to cross a room secured with laser beams. In order to do this, Wayne shoots a rope across the room and pulls on the other end, while Kyle balances on the taut rope.

Tasks

1) Look at the screenshot above and estimate Kyles weight force and the angle to which the rope spreads due to her weight.

2) Solve graphically: With what force F does Wayne have to pull in order to hold Kyle in this manner?

3) Solve graphically and compare: Which force F does Wayne have to muster up?

4) Evaluate: Is this type of break in realistic?

5) Which angle do you find if Bruce Wayne is able to pull with F = 600 N?

SOLUTION

1) Kyle must weigh around 50kg which equals 500 Newton. The rope spreads at an angle of around 170°.

2) Graphic Solution

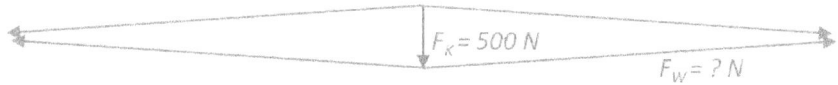

Parallelogram of forces drawn in a sensible scale:

3) Mathematical solution aided by trigonometry:

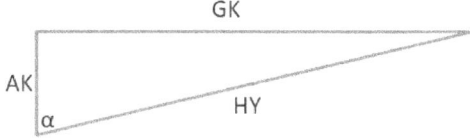

It holds that: $\alpha = \frac{170}{2} = 85°$, it follows that: $\cos(\alpha) = \frac{AK}{HY}$

therefore: $HY = \frac{AK}{\cos(\alpha)} = \frac{250}{\cos(85°)} = 2868N$ which corresponds to a weight of 186kg.

4) Bruce would never be able to muster up this kind of strength while standing up.

5) Again, with the help of trigonometry it is found that: $\cos(\alpha) = \frac{AK}{HY} = \frac{250}{600} = \frac{5}{12}$. From this it follows that there is an opening angle of $2 \cdot 65° = 130°$.

STAR WARS
THE FORCE AWAKENS

Keywords: Light, Mass, Quantum physics
Time of movie excerpt: see Youtube-Link: 1:04 min
Movie Trailer: https://www.youtube.com/watch?v=sGbxmsDFVnE

Content

In the science fiction fairy tale 'Star Wars', imperial armies and rebels fight for the domination of a far away galaxy. This movie series includes epic battles in space, as well as on planets. Alongside the famous lightsabers characters also use so called 'blasters', canons/guns shooting laser beams. One scene in the film shows a number of spaceships attacking the imperial troops, whereby the soldiers are hit by laser beams and thrown back.

Tasks

1) In the excerpt it can be seen how stormtroopers are being hit by red laser beams and spectacularly thrown back by their force. Make appropriate assumptions and calculate the mass of the light projectile vie the impulse conservation.

2) Calculate the mass of a red photon and, furthermore, determine the number of light particles per shot from task 1). (Ignore the quantum mechanical contradiction; Photons fly with light speed, which is not true in this scene)

3) Determine the power of the laser weapon in watts, using the number of photons per shot (duration of a shot t = 0.1s) and compare the result with the most powerful power plant on earth to date (internet research).

4) In Star Wars the imperial troops usually use red light in their weapons, the rebels use blue. What is smarter? Justify!

SOLUTION

1) Impulse conservation: $m_1 \cdot v_1 = m_2 \cdot v_2$

   Assumption: $m_{Soldier} = 90 kg$; $v_{Soldier} = 30 \frac{km}{h}$, $v_{Projectil} = 200 \frac{km}{h}$

   Thus it holds that: $m_2 = \frac{m_1 \cdot v_1}{v_2} = 13.5 \; kg$

   The glowing projectile (laser light) has a mass of 13.5kg

2) The mass of a photon can be calculated thus: $E = m \cdot c^2 = \hbar \cdot f$

   $m = \frac{\hbar \cdot f}{c^2} = 8.86 \cdot 10^{-28} kg$ (red light) or $1.44 \cdot 10^{-27} kg$ (blue light)

   Thus it holds that: $1.5 \cdot 10^{28}$ red photons or $9 \cdot 10^{27}$ blue photons per shot.

   Relation of 1 : 1.7.

   Simplified: Photons cannot fly with a speed of 200 kph (see above), they do not possess rest mass. However, evidently the Star Wars photons are able to do this.

3) The number of photons per shot can be calculated using the quotient of a photons power and energy, so $\frac{N_{Photon}}{t} = \frac{P}{E} = \frac{P}{\hbar \cdot f}$.

   $P = \frac{N_{Photon}}{t} \cdot E = \frac{N_{Photon}}{t} \cdot \hbar \cdot f = 37{,}768{,}200{,}000 \; W = 37 \; GW$ (red) or *22 GW* (blue).

   The most powerful atomic plants on earth reach a power of 1.5 GW. The strongest laser in the world reached the power of 2 million GW for a billionth of a second in 2015.

4) It is clearly more sensible to use blue laser weapons since the energy needed is lower then for red weapons.

DIE HARD

Keywords: Free fall, Velocity, potential energy, braking effort
Time of movie excerpt: 00:47:00 – 00:48:20
Movie Trailer: https://www.youtube.com/watch?v=2TQ-pOvl6Xo

Content

In the action movie 'Die Hard' from 1988 policeman John McClane fights against a group of kidnappers by himself in a closed off skyscraper.

In this scene McClane tries to escape from two of the kidnappers by hiding in an elevator shaft. He does this by lowering himself into the shaft on the strap of his machine gun. The strap comes lose and McClane falls into the depths. After about 13 meters of free fall he manages to hold on to a ledge and safe himself.

Tasks

1) What is John's velocity after falling down 13 meters?

2) How big is the force impacting John's fingers?

3) How can one imagine this force? Is the scene realistic?

SOLUTION

1) It applies: $s = \frac{1}{2} \cdot a \cdot t^2 \Leftrightarrow t = \sqrt{\frac{2 \cdot s}{a}} = 1.63s$ and $v = a \cdot t = 15.97\frac{m}{s} = 57.5\frac{km}{h}$.

2) If John McClane would have to hold onto the ledge with the tips of his fingers, his fingers have to undertake a 'brake effort'. The braking distance his fingers have to perform cannot be longer than 1cm and since the potential energy $E_{pot} = m \cdot g \cdot h$ is transformed into breaking energy $E = F \cdot s$, under the assumption that McClane weighs roughly 80kg, it applies:

$$m \cdot g \cdot h = F \cdot s \quad \Leftrightarrow \quad F = \frac{m \cdot g \cdot h}{s} = 1,020,240 \ N$$

1) A breaking force of about 1 million Newton corresponds to the weight of a mass of about 100 tons. This means that John McClane would have to be able to lift a passenger aircraft with only the tips of his fingers. This is obviously impossible.

HOME ALONE

Keywords: Free fall, Acceleration, Gravitation
Time of movie excerpt: 00:40:00-00:56:00
Movie Trailer: https://www.youtube.com/watch?v=ddXUQu9RC4U

Content

The film 'Home Alone' from 1990 is about 8-year-old Kevin who lives with his parents and four siblings in a suburb of Chicago. Mistakenly left alone at home over Christmas, he has to defend the house against two burglars. This film, as well as its sequel 'Home Alone in New York' (1992) are said to be leading actor Macaulay Culkin's biggest successes which has not been able to follow since then. This particular scene shows burglar Marvin, sneaking into the house through the cellar. In the middle of the room he pulls on an old fashioned cord switch in order to turn on the lights. Instead of the light he is hit by an electric iron falling down through the laundry chute. In the film, the iron falls for about 3.5 seconds.

Tasks

1) Watch the excerpt closely and measure the time the iron takes to fall on Marvin's head. Calculate the height of the house using the measured time. Does this seem realistic?

2) From an earlier scene it can be estimated that the house, from cellar to attic, is about 13 meters in height. Calculate the actual time the iron would take to fall from the top.

3) How tall would the house be on the moon/Jupiter, if the falling time in the film was correct? $(a_{Moon} = 1.6\frac{m}{s^2} , a_{Jupiter} = 23\frac{m}{s^2})$

4) Calculate a realistic falling time using a given house height (13m) on the moon as well as on Jupiter.

5) On which celestial body does the film verifiably take place?

SOLUTION

It applies: $s = \frac{1}{2} \cdot a \cdot t^2$ or $t = \sqrt{\frac{2 \cdot s}{a}}$.

1) $t_{measured} = 3.5s \rightarrow s = 60m$

2) $t_{realistic} = 1.62s$

3) $s_{Moon} = 9.8m$, $s_{Jupiter} = 140m$

4) $t_{Moon} = 4s$, $t_{Jupiter} = 1.06s$

5) The film takes place on the moon.

STAR TREK
THE NEXT GENERATION

Keywords: Star Trek, Heisenberg Uncertainty Principle, Quantum Physics, Beaming, Transporter
Time of movie excerpt: Star Trek TGN, 6x02, 17:00 – 17:50
Background: https://www.youtube.com/watch?v=sysxnM279X0

Content

The television series 'Star Trek – The Next Generation' (1987-1994) chronicles the adventures of a crew on board a spaceship. In order to save money while shooting, teleportation ('beaming') was introduced to get characters from place to place – this way no space shuttles had to land on planets all the time.

The mentioned scene is from the second episode of season 6 and tells of the nervous engineer Reginald Barclay who makes some eerie observations during beaming. His colleague, chief O'Brien, promises to thoroughly check the transporter again, whereby he mentions a number of the transporters components.

Tasks

1) In his speech O'Brien mentions so called *'Heisenberg-Compensators'*. Research on the internet and speculate why these compensators were mentioned in a script from 1992. What could be their purpose?

2) In the course of a theoretical piece of work in the year 1993, a team of researchers around Charles Bennett found out how to bypass Heisenberg's uncertainty principle. Carry out a second online research in order to find information about the so called 'Einstein-Podolsky-Rosen paradox', and explain why the 'Heisenberg-Compensator' would not be necessary anymore nowadays.

SOLUTION

1) Heisenberg's uncertainty principle states that it is impossible to determine an atom's location and momentum at the same time. It is therefore not possible to disassemble a body molecule by molecule, and then put it back together again.

2) EPR paradox describes 'entangled' atoms. This means, that two atoms can be linked in a way that they behave exactly the same. In this way, in 1997, a light particle was 'beamed' from one location to another using quantum teleportation.

VAN HELSING

Keywords: Parallelogram of forces, Vampires
Time of movie excerpt: see Trailer
Movie Trailer: https://www.youtube.com/watch?v=3fdRKme00uI

Content

In the mystery-horror film 'Van Helsing', the hero of the same name, played by Hugh Jackman, is on the hunt for vampires and the undead. He gets employed by the catholic church to protect the life of a princess and help her defeat Count Dracula who is terrorising her country. Traditionally vampires are killed by a wooden stake through the heart.

Tasks

Using the same force of a hammer blow: Which stake would be more likely to penetrate a coffin lid? Copy the drawing into your notebook and complete the parallelogram of forces

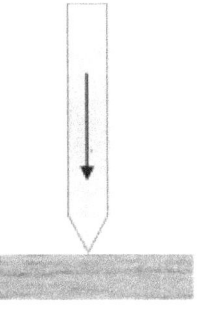
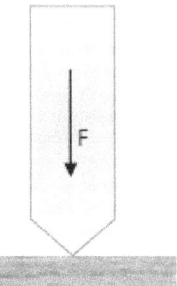

SOLUTION

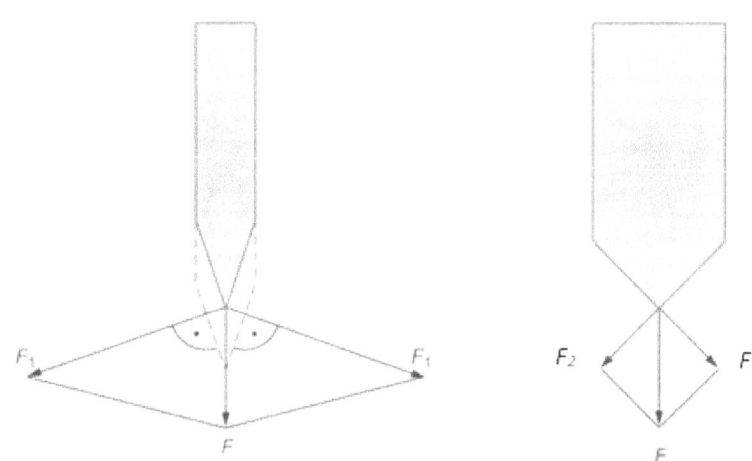

The two lateral forces F_1 and F_2, which stand perpendicular to the wedge-shaped surface, comprise, according to the law of parallelograms, the total force F.

As you can see, with the given F, both of the resulting forces F_1 and F_2 are bigger in the left picture.

HONEY, ...

Keywords: Shrinking, Volume, Mass
Time of movie excerpt: see Trailer
Movie Trailer: https://www.youtube.com/watch?v=_av5kqcMVm4&t=10s

Content

In the classic film 'Honey, I shrunk the kids' from 1989, the eccentric scientist Wayne Szalinski invents a machine with which he can shrink objects. One day he accidentally shrinks his own, as well as two neighbourhood children who suddenly find themselves in the garden surrounded by blades of grass that seem to be 10 meters tall. The children have to make it through a number of adventures and dangers to make it back to the laboratory.

Tasks

1) How big and tall is an average 14-year-old girl? Make appropriate assumptions.

2) An ant is about 3mm in length and weighs about 5 milligrams.

   On the basis of the above pictures, how tall and heavy will the girl be after having been shrunk?

3) Which other realistic biological, chemical and physical issues would the shrunk children be faced with?

SOLUTION

1) An average 14-year-old girl is about 1.55m/5'1'' tall and 50kg/110lbs heavy.

2) The girl would be about 2mm tall. This corresponds to a reduction ratio of 1:775. The weight follows from the volume, which changes to the 3rd power. Accordingly, the girl would get lighter by a factor of $4.6 \cdot 10^7$ and weigh about 0.1 milligram.

3) Different aspects contradict the shrinkage:

   • The air molecules are to big for the children to breathe.

   • The children would not be able to take in any food or water and would therefore die of dehydration after 3 days.

   • A small gust of wind would be enough to just blow them away.

   • They would be blind, because the eye would not longer be able to perceive anything.

The Physics of Hollywood

KING KONG

Keywords: King Kong, Weight, Mass, Volume, Magnification
Time of movie excerpt: see Trailer
Movie Trailer: https://www.youtube.com/watch?v=AYaTCPbYGdk

Content

King Kong was the first monster created for film and not adapted from literature. The Original was produced in 1933, in 2005 a remake was published, created by 'Lord of the Rings' director Peter Jackson. In reality a gorilla stands at about 1.5m in height, male gorillas can weigh up to 200kg.

Tasks

1) Estimating from the trailer - about how tall is the gorilla?

2) To what factor have height, width, and depth changed?

3) Give King Kong's weight. To what factor has it changed?

4) A body is supported by its skeleton. How much weight a bone can carry depends on the cross section of the bone. To what factor has the cross section of King Kong's bones changed in relation to a normal gorilla?

5) What does this mean for King Kong in reality?

SOLUTION

1) About 12 meters.

2) $\frac{12}{1.5} = 8$. The factor is 8.

3) For the volume it applies: length ·width · depth.
    $$8^3 \cdot 200kg = 512 \cdot 200kg = 102{,}400kg$$

4) Surface = length · width ➔ $8 \cdot 8 = 64$

5) He would not be able to stand.

MATRIX RELOADED

Keywords: Matrix, Acceleration, Energy, Perpetual Mobile
Time of movie excerpt: 00:10:25 – 00:10:52
Movie Trailer: https://www.youtube.com/watch?v=kYzz0FSgpSU

Content

In the science fiction film 'Matrix' almost all humans have been enslaved by intelligent machines, and are being kept dormant in giant breeding and storage facilities. They are hooked up to a highly complex computer simulation, the 'Matrix'. Since the sun is no longer able to penetrate the atmosphere, the machines are using the human body heat as their source of energy. The 'Chosen One' Neo is part of a group of rebels and realizes life is not just the simulation, but also has command over the Matrix and can change it. Within a second he can jump over a 40-storey high skyscraper.

Tasks

1) Physically reason why human bodies cannot be used as sources of energy. Write an energy flow chart for this purpose. (Energy from...to...towards)

2) Which acceleration does Neo reach while jumping over the aforementioned skyscraper?

3) How many multiples of the gravity acceleration is this?

SOLUTION

1) The film Matrix gives an example for a perpetual mobile. Humans cannot feed off of themselves. Furthermore, energy is not just lost there – the machines also use up energy.

   Energy flow chart:

   Energy from the sun → Photosynthesis → Energy stored as sugar within plants → animals eat the plants → food chain...

2) $s = 100m; t = 1s$

$$s = \frac{1}{2}at^2 \rightarrow a = 200\frac{m}{s^2}$$

3) This is 20 times the gravity acceleration.

SUPERMAN

Keywords: Superman, Acceleration, Velocity

Content

In 1938 the first comic featuring hero *Superman* was published. In the beginning, Superman was not able to fly yet but could jump incredibly high – one hop and he was on top of a 20-storey high building.

Tasks

1) The height of the jump is immediately linked to the take-off velocity. Derive the formula for this: $v^2 = 2 \cdot g \cdot h$

2) How does the height change, if the velocity is doubled? What is this correlation called?

3) How much take-off velocity does Superman need, to perfectly land on the aforementioned building? Make appropriate assumptions.

4) In order to get to this speed Superman need $^1/_{10}$ sec. Calculate the acceleration.

5) What take-off velocity do you need to jump on a chair?

6) In order to get to that speed, you need ½ of a second. Calculate the acceleration.

SOLUTION

1) $h = \frac{v^2}{2g}$

2) The height grow quadratically. There is a quadratic relationship.

3) Assumption: The building is about 60 meters high. Therefore, it applies:

$$v^2 = 2gh = 2 \cdot 9.81\frac{m}{s^2} \cdot 60m = 1177.2\frac{m^2}{s^2}$$

$$v = 34.31\frac{m}{s} = 123.5\frac{km}{h}$$

4) $a = \frac{v}{t} = \frac{34.41\frac{m}{s}}{0.1s} = 343.1\frac{m}{s^2}$

5) Assumption: The chair is 50 centimeters high.

$$v^2 = 2gh = 2 \cdot 9.81\frac{m}{s^2} \cdot 0.5m = 9.81\frac{m^2}{s^2}$$

$$v = 3.13\frac{m}{s}$$

6) $a = \frac{v}{t} = \frac{3.13\frac{m}{s}}{0.5s} = 6.23\frac{m}{s^2}$

WORLD WAR Z

Keywords: Airbag, Acceleration, Force, Car
Time of movie excerpt: 00:07:11 - 00:08:05
Movie Trailer: https://www.youtube.com/watch?v=Md6Dvxdr0AQ

Content

World War Z is an American Action film from 2013, which focuses on a world ruled by the undead. The beginning of the film shows the outbreak of an epidemic and the following panic. Gerry Lane (Brad Pitt) tries to escape from the city in his car at breakneck speed through the overcrowded streets, until he gets into a rear-end collision, which triggers the car's airbag to explode.

In normal traffic, rear-end collisions are a fairly regular occurrence, since drivers often use their mobile phones while driving. Hereby, the airbag explodes and throws the phone back onto the driver's face, which can cause serious injuries.

Tasks

Calculate the force, with which a phone would hit a driver in the face when accelerated by an airbag. Make assumptions about the different values and write them down neatly. An airbag takes about one twentieth of a second to inflate completely.

SOLUTION

Given: $t = 0.05\ s;\ m = 0.2\ kg;\ s = 0.3\ m$

Wanted: F = ?

Calculation: $F = m \cdot a$

from rest, $s = \frac{1}{2}at^2 \rightarrow a = \frac{2 \cdot s}{t^2}$

$$F = m \cdot \frac{2 \cdot s}{t^2} = 48\ N$$

This corresponds to a mass of about 5kg which can cause severe injuries to the face.

BIBI & TINA

Keywords: Horses, Velocity
Time of movie excerpt: individual episodes available on Youtube
Movie Trailer: https://www.youtube.com/watch?v=fNymKSIiU4M

Content

Bibi and Tina are the main characters in a famous german children's audio play series of the same name. The episodes about life on the horse ranch owned by Tina's mother are a spin-off of the series 'Bibi Blocksberg', which tells the tales of Bibi, a witch girl, and her adventures in magic. Especially the cartoon television series about Bibi and Tina is very popular and includes 85 episodes thus far. Almost every episodes begins or ends with a horse race between the two main characters.

Bibi's horse, Sabrina, can cover a quarter of a mile in 20.8 seconds when in full gallop. Tina's gelding, Amadeus, can trot one mile (1690 meters) in 1 min 53.4 sec.

Tasks

1) To what percentage is the speed of the gallop greater than that of the trot?

2) How many meters ahead is a galloping horse compared to a trotting horse after 500 meters, if both horses are half as fast as the respective film horse?

SOLUTION

1) Sabrina needs $4 \cdot 20.8 = 83.2$ seconds for one mile. Amadeus needs 113.4 seconds. Therefore, Amadeus needs 36% more time than Sabrina, or Sabrina needs 26% less time than Amadeus.

Calculation: $Sabrina \cdot x = Amadeus$ or $Amadeus \cdot y = Sabrina$

2) First of all, we have to calculate the velocities:

$$v_{Sabrina} = \frac{s}{t} = \frac{1609m}{83.2s} = 19.3\frac{m}{s}$$

$$v_{Amadeus} = \frac{s}{t} = \frac{1609m}{113.4s} = 14.2\frac{m}{s}$$

These velocities are then halved:

$$v_1 = 9.65\frac{m}{s} \text{ and } v_2 = 7.1\frac{m}{s}.$$

In the next step, we will check how much time the faster horse will take to run 500 meters:

$$t = \frac{s}{v} = \frac{500m}{9.65\frac{m}{s}} = 51.8s$$

The faster horse takes about 52 seconds to run the distance. The question was how far behind the slower horse would be in the same amount of time. Therefore, it applies:

$$s = v \cdot t = 7.1 \cdot 51.8 = 367.8$$

This means, the slower horse has only managed 367.8 meters and is therefore

500m – 376.8m = 132.2m behind.

JIM BUTTON

Keywords: Perpetual Mobile, Magnetism
Time of movie excerpt: Jim Button & the Wild 13, episode 2, 6:15 – 8:10
Movie Trailer: https://www.youtube.com/watch?v=y8p2ebyJp6s

Content

'Jim Button and Luke the Engine Driver' is a children's book by famous German author Michael Ende, published in 1960. The book was given a very popular adaption by the 'Augsburger Puppenkiste' an acclaimed German marionette theatre. The series was first shot in black and white in 1961/62, and completely re-shot in colour in 1976/77. All characters are depicted by puppets.

Tasks

In one of their adventures, Jim and Luke are stuck on a magnetic island. They modify the locomotive so that it is pulled by a magnet, just like a donkey following a carrot. Would it be possible to modify a rowing boat in a similar way (see above)? One could put a big magnet at the bow of the boat and a large iron rod at the stern. The magnet would constantly draw the piece of iron forward and power the boat. Would this kind of propulsion work? Justify your answer. Create a drawing and sketch the forces as part of your illustration.

SOLUTION

No, obviously this would not work. The key lies in Newton's laws: Every force creates a similar counterforce. In the same way the iron rod would push the boat forwards, the magnet would pull it back.

The Physics of Hollywood

ROBIN HOOD

Keywords: Oblique Throw, Velocity
Time of movie excerpt: 00:41:00 – 00:41:40 (Disney Version)
YouTube: https://www.youtube.com/watch?v=QLhYSw67pdg

Content

Robin Hood is the hero of many late medieval English ballades, which, over the centuries, have slowly morphed into the legends of the hero we know today. The oldest literature sources from mid 15th century still depict Robin Hood as a dangerous highwayman from humble origins, who preferably robbed greedy clergymen and aristocrats. Later he is described more and more positively. There have been numerous film adaptions, which particularly emphasis Robin Hood's shooting skills, such as the 1973 Disney version.

Tasks

1) The sheriff of Nottingham shoots an arrow with $v_0 = 56.6 \frac{m}{s}$ under an angle of 65° onto the finish line in 250m distance. Robin Hood shoots his arrow with an initial velocity of $v_0 = 52 \frac{m}{s}$ directly towards the target. Both shoot at the same time. Is it possible for Robin Hood's arrow to reach the target before the sheriff's arrow? Justify by using the velocity component in the direction of x.

2) Robin's arrow may pass the finish line in the air, but definitely has to reach it. Using the same initial velocity as in 1), which angle can he choose – at maximum – in order for his arrow to still reach the target line first?

SOLUTION

1) It applies: $t_{Robin} = \frac{s}{v} = \frac{250m}{52\frac{m}{s}} = 4.8s$

Therefore, Robin's arrow takes 4.8 seconds to target.

For the sheriff's arrow, we have to divide into x- and y-components where it applies:

$v_x = v_0 \cdot \cos(\alpha)$ and $v_y = v_0 \cdot \sin(\alpha) - g \cdot t$

Thus it follows: $v_x = 56.6\frac{m}{s} \cdot \cos(65°) = 23.9\frac{m}{s}$ and equals a flight time of 10.4 seconds for the sheriff's arrow.

2) Given is a flight time of 10.4 seconds maximum for Robin's arrow, in order to beat the sheriff. Therefore, it follows a minimum velocity of $v_R = \frac{s}{t} = \frac{250m}{10.4s} = 24\frac{m}{s}$. Using this we can now calculate the gradient angle because it applies

$$\cos(\alpha) = \frac{v_x}{v_0} = \frac{24\frac{m}{s}}{52\frac{m}{s}} = 0.46$$

from which follows an angle of $\alpha = 62.5°$.

MISSION IMPOSSIBLE 2

Keywords: Inelastic Collision, Momentum, Velocity, Braking Work
Time of movie excerpt: 01:44:30 - 01:45:30
Movie excerpt: https://www.youtube.com/watch?v=K2oKEqtQFyc (from 6:50)

Content

In the action film series 'Mission Impossible', Tom Cruise embodies a secret agent who stays in mind predominantly due to his spectacular stunts. During the showdown of Mission Impossible 2, there is a wild pursuit and two motorbikes purposely get into a head-on collision. From the scene it is evident that it is a completely inelastic collision. Both riders survive the crash and kind of 'stick' together after the impact.

Tasks

1) Calculate the resultant velocity with which the two motorcycles collide. Make appropriate assumptions.

2) Calculate the kinetic energy that occurs during collision. Make appropriate assumptions.

3) Assume the breaking distance is 20 centimetres. Calculate the force impacting the two men.

4) What weight does this force correspond with?

SOLUTION

1) During a completely inelastic collision, the two bodies 'morph' into one, so they end up with the same velocity v'. In order to calculate v' one needs the conservation of momentum, according to which the sum of the momentum before and after the collision has to be the same:

$$m_1 \cdot v_1 + m_2 \cdot v_2 = (m_1 + m_2) \cdot v'$$

Solved for v' results in:

$$v' = \frac{m_1 \cdot v_1 + m_2 \cdot v_2}{m_1 + m_2}$$

An appropriate assumption would be *80kg* of mass and *80kph* per driver. From this it follows a velocity of $v' = 80 \frac{km}{h}$.

2) $E_{kin} = \frac{1}{2} \cdot m \cdot v^2 = 19,753 \, J$

3) During the braking process, the body's whole kinetic energy has to be cancelled out through use of a force. This can be determined using $W = F \cdot s$.

$$F = \frac{W}{s} = \frac{19,753 \, J}{0.2m} = 98,765 \, N$$

4) This force, on earth, corresponds to a weight of about 10 tons. This stunt would be lethal in reality.

SPIDER-MAN 3

Keywords: Spiderman, Spider Silk, Traction, Force
Time of movie excerpt: 01:44:30-1:46:00
Movie Trailer: https://www.youtube.com/watch?v=MTIP-Ih_GR0

Content

In Spiderman 3 Peter Parker fights his nemesis Venom, the link to photography rival Eddie Brock and an extra-terrestrial symbiote. Venom, in his black suit, seems to be a bigger, stronger version of Spiderman. The third Spiderman film was the last one featuring Tobey Maguire as lead character and was the third most expensive film in history at the time. Director Raimi hide a number of friends and family in the film, including his three children.

Tasks

1) Spider silk is extremely tensile. Tensile strength describes the force per surface at which a type of material rips, this is measured in Pascal ($Pa = \frac{N}{m^2}$). The tensile strength of spider silk is at the magnitude of about 10^9 Pascal. Is it possible for a thread of spider silk with a diameter of *d = 1mm* to hold Spiderman? Make appropriate assumptions and justify.

2) During the film a police car is thrown into the air and caught by spider web just as it is about to hit a group of innocent civilians. Would this be possible? Make appropriate assumptions.

SOLUTION

1) A thread of spider silk with a diameter of 1 millimetre, has a cross section of $A = \Pi \cdot r^2 = 0.79 \ mm^2$. Spider silk possesses a tensile strength of about 10^9 Newton per square meter. Using the rule of three it can be found that spider silk displays a tensile strength of 1000 Newton per square millimetre or 790 Newton for 0.79 mm^2. This corresponds to about 80kg.

2) For a car with a weight of 1,200kg, it would be enough to use a spider web made up of 15 threads. If Spiderman was able to produce threads double as thick, only 4 threads would be enough to support the car. The Spider silk threads shown in the film are likely even thicker – the scene is therefore quite realistic.

JAMES BOND CASINO ROYALE

Keywords: Free Fall, Acceleration, Force, Braking Effort
Time of movie excerpt: 00:12:20 – 00:14:20
Movie Trailer: https://www.youtube.com/watch?v=36mnx8dBbGE

Content

James Bond, agent 007, is a secret agent, thought up by Ian Fleming, who works for British secret service MI6. Numerous film versions with changing lead actors have been published since 1962. In 'Casino Royal' from 2006, Bond chases bomb manufacturer Mollaka. On a building site protagonist and antagonist climb up crane towers and scaffolding, run across wonky surfaces, climb up ropes and fight on top of crane booms. In one instance, Bond jumps from one crane to the next and falls about 5 meters before catching himself by his arms.

Tasks

1) How large is Bond vertical velocity as he catches himself?

2) What force affects Bond arms estimating a braking distance of about 10cm?

3) What mass does this strain correspond to?

SOLUTION

1) We are talking about an accelerated movement; therefore, it applies $v = a \cdot t$ with a falling time $\left(s = \frac{1}{2} \cdot a \cdot t^2\right)$ $t = \sqrt{\frac{2 \cdot s}{a}}$ so $v = a \cdot \sqrt{\frac{2 \cdot s}{a}} = 9.9 \frac{m}{s}$ or $35.6 \frac{km}{h}$.

2) During the braking process, the body's whole kinetic energy has to be cancelled out by a force. This can be calculated using $W = F \cdot s$. Assuming the agent's weight to be about 80kg it applies:

$$F = \frac{W}{s} = \frac{E_k}{s} = \frac{\frac{1}{2} \cdot m \cdot v^2}{s} = \frac{\frac{1}{2} \cdot 80 \, kg \cdot \left(9.9 \frac{m}{s}\right)^2}{0.10 \, m} = 39{,}204 \, N$$

3) This corresponds to a weight of roughly 4 tons. It would be impossible to catch oneself after such a jump, using just one's arms.

The Physics of Hollywood

ERASER

Keywords: Recoil, Firearm, Momentum Conservation, Force, Speed of Light
Time of movie excerpt: 01:46:10 – 01:46:28
Movie Trailer: https://www.youtube.com/watch?v=31_OEhX30sY

Content

In the fiction action film 'Eraser', Arnold Schwarzenegger fights against traitors within the government. In one of the scenes he casually holds two so called 'Railguns' in wildly shoots everything around him. Such guns accelerate their ammunition to several kilometers per second. Under the assumption that Schwarzenegger's (invented) guns accelerate 100g heavy ammunition to 3000 m/s, answer the following questions.

Tasks

1) How much force, per shot, effects Arnolds arms?

2) Is it possible to effectively fire this weapon?

3) Actio est reactio (action is reaction). Using momentum conservation, determine the velocity with which Schwarzenegger and the guns are thrown backwards with every shot.

4) In the film it is said that the projectiles fly with 'nearly speed of light'. Comment.

SOLUTION

1) The reaction force F_R corresponds to the acceleration force F_A, which the projectile and the forward surging gasses are exposed to. The acceleration force is the product of projectile mass m_G and projectile acceleration a:

$$F_R = F_A = m_G \cdot a$$

Idealised, the projectile acceleration can be calculated using the muzzle velocity v_0 of the projectile, and the weapon's barrel length s.

$$a = \frac{v_0^2}{2 \cdot s}$$

With an estimated barrel length of 60cm this means a force of $F_R = m_G \cdot \frac{v_0^2}{2 \cdot s} = 750,000\ N$ which equates to roughly $76,000\ tons$.

2) No. It is not possible to fire this weapon.

3) The momentum conservation states that Schwarzenegger, along with his weapon, experiences the same negative momentum as the projectiles, so

$$m_1 \cdot v_1 = m_2 \cdot v_2$$

Under the condition that Schwarzenegger plus his weapon weight about 130kg, the result, solved for v_2 is:

$$v_2 = \frac{m_1 \cdot v_1}{m_2} = \frac{0.1\ kg \cdot 3000\frac{m}{s}}{130\ kg} = 2.3\frac{m}{s} = 8.3\frac{km}{h}$$

Considering the large amount of shots fired, it is rather unlikely that he is firing off the cuff like this.

4) The projectile velocity is, according to our assumptions, $3000\frac{m}{s}$. Speed of light, however, is $300,000,000\frac{m}{s}$, and thus larger by a factor of 10^5. Even with the ludicrous speed based on our assumptions it would be impossible to hold the weapon – increasing the velocity even further, would be absolutely impossible.

The Physics of Hollywood

SPIDER-MAN 3

Keywords: Mechanics, Freefall, Spiderman, Braking Effort
Time of movie excerpt: 01:46:10 – 01:46:28
Movie Trailer: https://www.youtube.com/watch?v=ukGsu63s2dk

Content

In the comic series 'Spider-Man', young Peter Parker develops superhuman skills after having been bitten by a radioactive spider. Dressed in a red costume he fights as 'Spiderman' for the weak. In the third and last Spiderman film featuring lead actor Tobey Maguire from 2007, Peter Parker's girlfriend Mary Jane falls for about 19 seconds during the film's finale, until the hero manages to save her in the last second.

Tasks

1) How fast are Parker and Mary Jane at the end of the fall?

2) From the horrible medieval punishment practice of quartering, we know that an arm can withstand a strain of about 3,000 Newton before it rips off. Can Spiderman withstand the strain, if we estimate the braking distance to start at 20 meters?

3) How thick does the thread of spider silk have to be at a minimum to support Parker and Mary Jane?

4) Look at the task, disregarding the time, and assuming that Spiderman is falling 80 meters.

Hint: A thread of spider silk is known to have a tensile strength of about 1 Gigapascal or 10^9 Newton per square meter.

SOLUTION

1) Mary Jane falls for about 19 seconds. Disregarding the air resistance, this results in:

$$= \frac{\Delta v}{t} \Leftrightarrow v = a \cdot t = 9.81 \frac{m}{s^2} \cdot 19s = 186.39 \frac{m}{s} \sim 671 \frac{km}{h}$$

2) The achieved 'braking effort' has to cancel out the total kinetic energy. The kinetic energy (at an estimated mass of 130kg for Spiderman and Mary Jane) amounts to:

$$E_k = \frac{1}{2} \cdot m \cdot v^2 = \frac{1}{2} \cdot 130kg \cdot \left(186.39 \frac{m}{s}\right)^2 = 2,258,180\,J$$

The braking effort $W_{Break} = F \cdot s$ is exactly as great as the kinetic energy. Solved for F it applies: $= \frac{W_{Break}}{s} = \frac{E_k}{s} = 110\,kN$

Therefore, a force of 110,000 N is applied to Spiderman's arm, which means that him and Mary Jane would keep plummeting into the depth, while only his bloody arm would remain dangling from the spider thread.

3) The spider thread would have to be $\frac{110,000\,N}{1,000,000,000 \frac{N}{m^2}} = 0.00011m^2$ thick. Assuming the thread is round, it would apply: $\pi \cdot r^2 = 0.00011m^2$, from which follows that the thread would have to have a radius of 0.59 cm.

4) Assuming Mary Jane would 'only' drop down about 80 meters it would apply: $v = \sqrt{2 \cdot g \cdot s} = 40 \frac{m}{s} \sim 140 \frac{km}{h}$.

It follows: $E_k = \frac{1}{2} \cdot m \cdot v^2 = 102\,kN$ - compared to the braking effort it would result a necessary force of 5,100 N. Here, too, his arm would be history.

LIVE FREE OR DIE HARD

Keywords: Recoil, Firearm, Momentum Conservation, Force
Time of movie excerpt: 01:59:20 – 2:00:10
Movie Trailer: https://www.youtube.com/watch?v=8Jz-8UcCiws

Content

In Die Hard 4.0 Bruce Willis once more gets into the role of the savior of the world. In a large storage facility, a showdown between good and evil takes place. One of the bad guys fires his weapon precisely towards the hero hiding behind barrel, holding the gun in his outstretched hand and perfectly still. Despite the recoil his hand does not move.

Tasks

Is it possible to fire shots from a gun one handed, without the arm noticeably moving due to the recoil? Make appropriate assumptions.

Mass of projectile: 7.5g

Initial velocity: 350 m/s

Barrel length: 12 cm

SOLUTION

The reaction F_R corresponds to the acceleration force F_A, which the projectile is exposed too. The acceleration force is the product of the projectile mass m_G and the projectile acceleration a:

$$F_R = F_A = m_G \cdot a$$

Idealised, the projectile acceleration can be calculated using the muzzle velocity v_0 of the projectile and the weapon's barrel length:

$$a = \frac{v_0^2}{2 \cdot s}$$

With a barrel length of 12cm this results in a force of $F_R = m_G \cdot \frac{v_0^2}{2 \cdot s} = 3{,}828\ N$, which equates to a mass of about 390 kilograms.

On YouTube there are a number of compilations of recoil videos (see QR-code). They illustrate very well that physical calculations (despite simplification and disregarding of a recoil shock absorber) are closer to reality than the scenes shown in the film.

TERMINATOR 2

Keywords: Motorbike, Jump, Rotation (Movement), Angular Velocity, Free Fall
Time of movie excerpt: see YouTube link
Movie Trailer: https://www.youtube.com/watch?v=o_TNfLpc91M

Content

Arnold Schwarzenegger chases the evil Terminator on a motorbike. Hereby, Schwarzenegger drives straight towards a pointed concrete edge, which separates the canal below into two different directions. With undiminished speed, Arnold jumps – on his motorbike - down into the canal bed which lies approximately 5 meters below. After the landing, the chase resumes immediately. From the scene in the film it is clear that the T-800's motorbike does not drive over a ramp or anything else of the sort when coming off the concrete edge. The front wheel, logically, leaves the edge before the back wheel. If the motorbike followed the physical rules of the free fall, it would dip forward and not as shown in the scene, retain a horizontal trajectory.

From watching the scene, we can estimate the driving speed to be about 60kph and the fall about 6m. The wheel base is assumed to be 1.7 meters.

Task

1) The front wheel leaves the concrete edge at the time t_0. Calculate the time t_1 at which the back wheel leaves the edge.

2) During this time the motorbike has already dipped forward. How many meters has it 'fallen'?

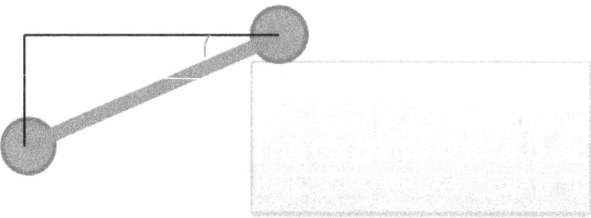

3) What angle now corresponds to the tilted position of the motorbike? (Sketch)

4) What angular velocity ω corresponds with the rotation of the motorbike?
To simplify we will assume that the angular velocity stays constant over time.

5) After which time t_{total} does the front wheel hit the ground?

6) What angle does the motorbike have now?

SOLUTION

1) For a constant motion it applies $t_1 = \frac{s}{v} = \frac{1.7m}{16.7\frac{m}{s}} = 0.1\ s$.

2) The front wheel experiences an acceleration motion downwards, which is independent of the motion in direction of x:

$$s = \frac{1}{2} \cdot a \cdot t^2 = \frac{1}{2} \cdot 9.81 \frac{m}{s^2} \cdot (0.1\ s)^2 = 0.05m$$

The front wheel has now dipped down by 5cm.

3) For the trigonometry it applies: $\sin(\alpha) = \frac{OPP}{HYP} = \frac{0.05}{1.7} = 0.03$ which results in an angle of 1.7°.

4) The angular velocity is therefore 17°/s

5) The front wheel takes exactly $t = \sqrt{\frac{2 \cdot s}{a}} = \sqrt{\frac{12\ m}{9.81\frac{m}{s^2}}} = 1.1s$ to fall down the 5 meters.

6) After one second, the motorbike exhibits a tilting angle of 17.2°. It definitely would not hit the ground with both wheels at the same time. The scene is, from a physical point of view, not correct.

INDEPENDENCE DAY

Keywords: Volume, Gravity, Mass
Time of movie excerpt: see YouTube link
Movie Trailer: https://www.youtube.com/watch?v=kA2WzBi2grE

Content

In the apocalyptic movie Independence Day, aliens attack the earth. During the film, viewers are told that the (approximately) semi-spherical mother ship has a diameter of about 550km and possesses a mass as big as ¼ of the moon. Throughout the film, smaller, disc shaped space ships attack cities, each ship has a diameter of 24km.

Tasks

1) Find the volume and the mean density of the mother ship.

2) Research: What is Uranus' density? Pay attention to the units! What can you deduce from this?

3) Find the mass of the attacking spaceships. Make appropriate assumptions.

4) With a radius of 6300km, the earth has a mass of $6 \cdot 10^{24}$ kg. Calculate the force with which earth and the mother ship attract each other

5) The moon has a mass $7 \cdot 10^{22}$ kg and is about 384000km far away. Calculate the force attracting moon to earth.

6) Compare 4) and 5). What does this mean for the people of New York?

SOLUTION

1) It is given that $r = 275\ 000m$ and $m = \frac{1}{4} \cdot 7 \cdot 10^{22}\ kg$

$$V_{Sphere} = \frac{4}{3} \cdot \pi \cdot r^3 \quad \rightarrow \quad V_{Mother\ Ship} = \frac{1}{2} \cdot \frac{4}{3} \cdot \pi \cdot r^3 = \frac{2}{3} \cdot \pi \cdot r^3$$

$$\rho_{Ship} = \frac{m}{V} = \frac{m}{\frac{2}{3} \cdot \pi \cdot r^3} = \frac{1 \cdot 7 \cdot 10^{22} kg}{4 \cdot \frac{2}{3} \cdot \pi \cdot (275{,}000m)^3} = 401{,}773\ \frac{kg}{m^3}$$

2) $\rho_{Uran} = 19.16 \frac{g}{cm^3} = 19160 \frac{kg}{m^3}$

3) Assumption: The ships are spherical discs with a height of 100 meters. At a radius of 12km it applies: $V = 2\pi r^2 \cdot h = 9 \cdot 10^{10} m^3$

It further applies: $m = \rho_{Ship} \cdot V = 401773 \frac{kg}{m^3} \cdot 9 \cdot 10^{10} m^3 = 3.6 \cdot 10^{16} kg$

4) It applies: $F = f \frac{m_1 m_2}{r^2} = 6.67 \cdot 10^{-11} \frac{6 \cdot 10^{24} \cdot 3.6 \cdot 10^{16}}{6{,}300{,}000^2} = 3.6 \cdot 10^{17} N$

5) It applies: $F = f \frac{m_1 m_2}{r^2} = 6.67 \cdot 10^{-11} \frac{6 \cdot 10^{24} \cdot 7 \cdot 10^{22}}{390{,}000{,}000^2} = 1.9 \cdot 10^{20} N$

6) The attraction force between earth and spaceship are weaker by a factor of about 500. Therefore, there would not be an affect on the tides due to the small attacking space ships.

YES MAN

Keywords: Free Fall, Kinetic Energy, Bungee
Time of movie excerpt: see YouTube link
Movie Trailer: https://www.youtube.com/watch?v=K9Id30IR5-Y

Content

'Yes Man' is an American comedy film from 2008, in which Jim Carry has the lead role. The film is based on an autobiographical book by Danny Wallace, a British author, producer and journalist, who spent one year of his life responding with 'yes' to every question and decision he faced and who then wrote down his experiences. In one scene, Carrey jumps off a bridge using a bungee rope. Carrey did not have a stunt double do this scene for him, but jumped himself. It was his first bungee jump and he liked the idea of this translating onto the film. Director Peyton purposely had this scene shot last, in case something would have happened.

Tasks

Jim Carrey is supposed to stop falling at 10m above the ground and there should never be more than 3g (three times the gravitational acceleration) impacting his body. How long is the rope at maximum? This task is easy to solve argumentatively, not so easy to solve physically. Start with your line of reasoning and then regard the situation from a physical point of view.

SOLUTION

Argumentatively it is easy to describe the situation: If Carrey falls with 1g, but can be decelerated with 3g, 'drop distance' and 'braking distance' behave in the same ratio. The ratio of rope to stretched rope would be 3:1.

Since the braking process is not supposed to exceed the velocity of 3g, we can also argue with the accelerated motion:

If $s = \frac{1}{2}at^2$ is the braking distance and for free fall it applies $t = \frac{v}{a}$, then

$$s = \frac{1}{2} \cdot a \cdot t^2 = \frac{1}{2} \cdot \frac{v^2}{a}$$

At the end of the free fall, the kinetic energy is just as large as the potential energy $m \cdot g \cdot h$ with 'h' being the length of the **none** stretched rope 'l':

$$\frac{1}{2} \cdot m \cdot v^2 = m \cdot g \cdot l$$
$$v = \sqrt{2 \cdot g \cdot l}$$

Insert above.

$$s = \frac{1}{2} \cdot \frac{v^2}{a} = \frac{1}{2} \cdot \frac{2 \cdot g \cdot l}{a} = \frac{g \cdot l}{a}$$
$$\frac{s}{l} = \frac{g}{a} = \frac{g}{3g}$$

Therefore, the ratio is 3:1 of rope to braking distance.

STAR TREK 7

Keywords: Gravity, Stars
Time of movie excerpt: 01:21:00 – 01:21:45
Movie Trailer: https://www.youtube.com/watch?v=MUieGh1fHSI

Content

In the film Star Trek Generations, villain Soran inflates the sun of a solar system into a supernova in order to change the gravitational pull within the system. With this changed gravitational pull, he is trying to redirect an energy band, the Nexus, onto a planet. In the film it is shown how a probe hits a star, which changes immediately and creates a shock wave, which, in turn, changes the course of the energy band. It reaches the planet, which, after a short period of time, is also hit by the shock wave and destroyed.

Tasks

1) Argue from a physical point of view: Why does the expansion of the sun into a supernova in the first instance not change the gravitational pull within the solar system?

2) At which point in time t_1 would such a change be noticeable?

SOLUTION

1) The gravity within the solar system depends on the mass of the sun within the system.
    However, through the expansion of the sun into a supernova, the mass does not change (pic. 2) - therefore, there are no effects on the gravitational conditions within the system.

2) Gravity only changes once a part of the sun's mass is on the other side of the planet (pic. 3).

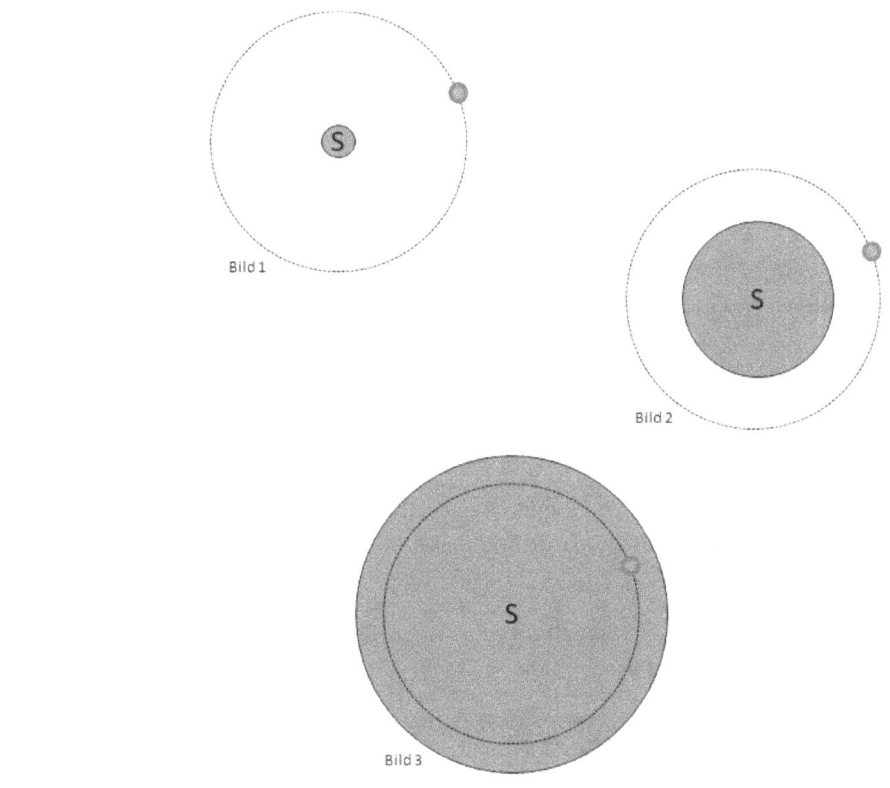

Bild 1

Bild 2

Bild 3

STAR TREK DEEP SPACE NINE

Keywords: Shrinking, Vibration, Pendulum
Time of movie excerpt: Star Trek DS9, 6x14
Movie Trailer: https://www.youtube.com/watch?v=xvIckMJkgKc

Content

Harmonic vibration/oscillation is a pillar of theoretical physics. The human vocal chords are not exactly thread pendulums, but simplified they could still be viewed as harmonic vibration (thread pendulum). In the series Star Trek Space Nine the crew comes across a space phenomenon, which shrinks part of the crew to 1/170 of their original size. In the episode it is mentioned that the air molecules outside of the shrunk spaceship are too big to breathe now – however, a different physical aspect is not mentioned.

Tasks

1) Which factor does the vibration period of a pendulum depend on?

2) The pitch (frequency) of the human voice lies at about 200 Hertz (vibrations per second). To what factor does the vibration period of the shrunken crew members' vocal chords change?

3) How does the frequency of speech change? With what pitch do the shrunken crew members speak now?

4) The human hearing ability lets us hear from 20 to 20 000 vibrations per second. Could we understand shrunk Captain Sisko?

5) Simplified, hearing ability depends on the surface of the ear drum. So, vice versa, can the shrunken crew still understand normal sized humans?

SOLUTION

1) For the vibration period of a pendulum it applies: $T = 2\pi\sqrt{\frac{l}{g}}$. Thus, the only changeable variable is l.

2) For the change in vibration period it applies: $T_{small} = 2\pi\sqrt{\frac{1/170 \, l}{g}}$.

   The changed factor is therefore $\sqrt{\frac{1}{170}} \sim \frac{1}{13}$. The vibration period T is this smaller by a factor of 13. (Note: When viewing the vocal cords as surfaces rather than lines, the factor would have to be squared!)

3) It applies: $T = \frac{1}{f}$. The frequency is thus larger by a factor of 13.

4) $200 Hz \cdot 13 = 2,600 \, Hz$. Yes, we could still understand the crew.

5) Simplified, the ear drum can be viewed as a surface – therefore, it shrinks at a factor of 13^2.

   Accordingly, the audible frequency increases by a factor of 13^2. The audible range would this lie between 3400Hz and 3400000Hz, meaning the crew could neither hear us, nor themselves.

MONSTERS VS. ALIENS

Keywords: Magnification, Mass
Time of movie excerpt: see YouTube link
Movie Trailer: https://www.youtube.com/watch?v=NiS1G2__DKc&t=14s

Content

In the animated movie Monsters vs Aliens, protagonist Susan (65kg) is hit by an asteroid on the day of her wedding and grows from a height of 1.66m to 15m. Together with a group of other strange creatures, she establishes a tight team in order to fight an alien invasion.

The weight that a bone can support depends on its cross section. For a woman of a similar build to Susan, the thigh bone will be able to withstand about 8,000 N a vertebra (which only has to carry half of the body weight) about 3,600N.

Tasks

1) Are the bones able to support Susan after she is magnified?

2) Which size would she be at maximum before her bones would break?

3) The pitch (frequency) of the human voice is about 200 vibrations per second. To what factor would Susan's vibration period change? (Hint: Look at the vocal chords as a thread pendulum)

4) The human hearing ability ranges from 20 to 20 000 vibrations. Would we still be able to understand Susan?

5) Simplified, hearing ability depends on the size of the eardrum (Hint: Surface!). So in reverse, would Susan be able to still understand normal sized humans?

SOLUTION

1) First of all, one has to establish the magnification factor x.
$$1.66 \cdot x = 15 \quad \rightarrow x = 9.03$$

This factor applies to length, width and depth. Susan's weight (volume), therefore changes with the third power:
$$65\text{kg} \cdot 9.03^3 = 47{,}958\text{kg}.$$

The bones' cross section (surface), however, only growth with the second power. The weight-bearing capacity is therefore:
$$8{,}000\text{N} \cdot 9.03^2 = 652{,}327\text{N} \approx 65\text{t}$$
$$3{,}600\text{N} \cdot 9.03^2 = 293{,}547\text{N} \approx 30\text{t}$$

Both bones are therefore able to support Susan's weight (note: The vertebrae only has to carry half the weight)

2) Wanted is the maximum magnification factor x.

It applies:

$$650\text{N} \cdot x^3 = 8000N \cdot x^2 \quad \text{or} \quad 325\text{N} \cdot x^3 = 3600N \cdot x^2$$

(The growing weight force can be no more than the maximum weight-bearing capacity)

It follows that $x_O = 12.3$ and $x_R = 11.1$

Therefore, Susan could be $1.66\text{m} \cdot 11.1 = 18.38\text{m}$ tall at maximum.

SOLUTION

3) For the vibration period of a pendulum it applies: $T = 2\pi\sqrt{\frac{l}{g}}$. The only changeable variable is therefore l. For the change in vibration period it applies:

$$T_{large} = 2\pi\sqrt{\frac{9 \cdot l}{g}}$$

The changed factor is therefore $\sqrt{9} \sim 3$. The vibration period T is this changed by a factor of 3, correspondingly, the frequency decreases by a factor of 3.

4) Susan talks at a frequency of $f = \frac{200}{3} = 66$ Hz. Yes, we can still understand her.

5) The eardrum (surface) grows by a factor of 9^2. This means that Susan will hear at a frequency of 1 – 246 Hz. (More realistically, the cut-off would be around 10Hz, even that would already be subsonic noise)

SPEED

Keywords: Angle Velocity, Jump, Freefall
Time of movie excerpt: see Youtube link
Movie Trailer: https://www.youtube.com/watch?v=dKJa-KQNjQU&t=60s

Content

In the movie Speed, terrorists place a bomb inside a bus, approximately 10 meters in length. It is supposed to explode once the driver as falls below a speed of 80khp (50mph). One of the policemen who helped foil the first terror attack, manages to get on the bus, which is speeding through Los Angeles. He does everything to keep up the speed. As the bus driver gets onto an empty highway, 40m in height, he realizes that he is heading towards and unfinished part of the road. Parts of the bridge are incomplete, there is a gap in the road about 15m wide. The bus driver accelerates to 110kph, jumps over the bridge and lands the bus safely on the other side.

Tasks

1) How long does it take the bus to jump the 15m?

2) The front axle leaves the concrete edge at a point in time t_0. Calculate the point in time t_1, at which the rear axle leaves the edge.

 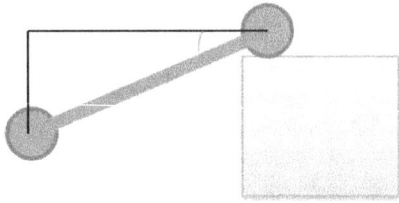

3) During this time the front axle has already dipped down. How many meters has it 'fallen' at this point? (Distance between axles is 8m)

4) What angle does the incline of the bus now correspond to? (see diagram)

5) What angle velocity does the bus's rotation now correspond to?

6) Where - in terms of the bus's height – will the approaching highway hit the bus?
   What does this mean for the driver?
   To simplify we assume that the angle velocity ω stays constant across time.

7) Assuming the second highway was not in the way, after what time t_3 would the bus hit the ground?

8) What angle would the bus have at that point?

SOLUTION

1) It applies: $t = \frac{s}{v} = \frac{15 \text{ m}}{30.6\frac{m}{s}} = 0.5$ s

2) If the distance between the axles is 8 meters, it applies: $t = \frac{s}{v} = \frac{10 \text{ m}}{30.6\frac{m}{s}} = 0.26$ s

3) Free fall = accelerated motion.

   It applies: $s = \frac{1}{2}gt^2 = \frac{1}{2} \cdot 9.81\frac{m}{s^2} \cdot (0.26s)^2 = 0.37$m

4) It applies: $\sin(\alpha) = \frac{leg\ (a)}{HY} = \frac{0.37\text{m}}{8\text{m}} \rightarrow \alpha = 2.65°$.

5) It applies: $\omega = \frac{\varphi}{t} = \frac{2.65°}{0.26s} = 10.2°\ s^{-1}$

6) Here we have to start at task 3): How many meters has the front of the bus fallen during the 15m jump?

   It applies: $s = \frac{1}{2}gt^2 = \frac{1}{2} \cdot 9.81\frac{m}{s^2} \cdot (0.5s)^2 = 1.23$m

7) Height of the highway: 40m

$$t_3 = \sqrt{\frac{2s}{g}} = 2.86s$$

8) It applies: : $\varphi = \omega \cdot t = 10.2°\ s^{-1} \cdot 2.86s = 29.1°$.

KNIGHT RIDER

Keywords: Velocity, Air Resistance, Density, Car
Time of movie excerpt: see YouTube link
Movie Trailer: https://www.youtube.com/watch?v=oNyXYPhnUIs

Content

In the television series 'Knight Rider' from the 80s, former cop Michael Knight and his talking car K.I.T.T. fight crime for the 'Foundation of Law and Government'. The car, K.I.T.T., was able to drive up to 480 kph. As a mobile service station, a truck was used, in which K.I.T.T. was repaired and transported.

Tasks

1) How much air does K.I.T.T. move when he drives one kilometer? And how much – in comparison - does the truck move? The density of air is about 1.204 kg/m3. Make assumptions for missing data.

SOLUTION

Air volume K.I.T.T.: $2m \cdot 1.5m \cdot 1000m = 3000m^3$.

Weight of air: $3000m^3 \cdot 1.204\frac{kg}{m^3} = 3612kg$

K.I.T.T. has to move 3 ½ tons of air per kilometer.

In comparison the truck: $2.5m \cdot 3.5m \cdot 1000m = 8750m^3$.

Weight of air: $8750m^3 \cdot 1.204\frac{kg}{m^3} = 10535kg$.

The truck has to move 10 ½ tons of air per kilometer. Over a distance from New York to Washington DC that would be 3800 tons!

The Physics of Hollywood

FAST & FURIOUS

Keywords: Car, Force, Velocity
Time of movie excerpt: 01:19:25 – 01:21:20
Movie Trailer: https://www.youtube.com/watch?v=T3J9yYnKLA0

Content

The Fast & Furious series contains a number of movies, in which several heroes drive their sports cars through spectacular action filled scenes. During the seventh movie, the protagonists jump from one skyscraper to the other in their car, while not wearing their seatbelts. Time after time people make the tragic error of thinking seatbelts are unnecessary, because it is supposedly possible to 'catch' oneself during a minor collision by bracing one's arms against the steering wheel.

It is possible to insert velocity (in kph) into the formula $h = v^2/254$ in order to find a distance. This can be viewed as the height, which one falls at top speed (i.e. from a table to the floor) in order to be able to imagine the kind of forces which taking place at such velocities.

Tasks

1) Derive the formula. Note the transformation of units.

2) A fall from what height does a collision at 30 kilometers per hour correspond to.

3) Think back to the motorbike crash from *Mission Impossible II*: What height of free fall does that collision correspond to?

SOLUTION

1) We start by equating the kinetic energy (within a travelling car) with the potential energy (of falling from a certain height).

$$E_{kin} = E_{pot}$$

$$\frac{1}{2}mv^2 = mgh$$

Since the car's velocity is usually given in kph, this has to be taken into account within the formula:

$$\frac{1}{2}m\left(\frac{v}{3.6}\right)^2 = mgh$$

$$\frac{1}{2}\left(\frac{v}{3.6}\right)^2 = gh$$

$$\frac{\frac{1}{2}v^2}{12.96} = gh$$

$$h = \frac{v^2}{2 \cdot 12.96 \cdot 9.81} = \frac{v^2}{254}$$

2) If $v = 30\frac{km}{h}$ it applies: $h = \frac{v^2}{254} = 3.54m$

To illustrate, this would correspond to jumping of the 3-meter diving board at the swimming pool. This cannot be caught using one's arms.

3) Assumption: Both motorbikes travel towards each other at 80 kph:

$$h = \frac{v^2}{254} = 25m$$

The collision would correspond to a fall at full speed from a height of 25 meters into asphalt.

FLASH

Keywords: Kinetic Energy, Velocity, Braking Effort, Speed of Light,
Time of movie excerpt: see YouTube link
Movie Trailer: https://www.youtube.com/watch?v=r9-DM9uBtVl&t=35s

Content

The superhero *Flash* debuted in a comic book in 1940, in which readers are told about Barry Allen, a scientific staff member of the police department. On the way home from work he stops at chemical laboratory, as a lightning bolt hits, and equips him with the ability to run at the speed of light. Except for his speed, Flash does not possess any other superpowers. In Flash #124, our hero stops a bullet, fired by an assassin, just a few centimeters from the president's chest. For a super fast hero, the speed of a bullet (7.5g and $350\frac{m}{s}$) is no problem – but what about catching it?

Tasks

1) Can Flash stop the bullet in time, if he manages to grab it 7cm away from the president's chest?

2) Assuming that Flash can lift a load of 50kg with one arm, how far away from the president's chest would he have to reach the bullet at the latest?

1)   First of all, we have to calculate the bullet's kinetic energy:

$$E_{kin} = \frac{1}{2}mv^2 = \frac{1}{2} \cdot 0.0075kg \cdot (350\frac{m}{s})^2 = 459.4\ J$$

Subsequently, we have to look at the 'braking effort': The total kinetic energy as to be decelerated, therefore it applies:

$$E_{kin} = W = F \cdot s$$

Since we are looking for the force F, it applies:

$$F = \frac{W}{s} = \frac{E_{kin}}{s} = \frac{459.4J}{0.07m} = 6562.5\ N$$

This corresponds to a weight of about 650kg. It is impossible to catch this kind of load with one hand.

1)   We can use the values from above. Instead of *s* we *F* as a given.

$$s = \frac{W}{F} = \frac{E_{kin}}{F} = \frac{459.4J}{500N} = 0.92m$$

INDEPENDENCE DAY

Keywords: Gravity, Force
Time of movie excerpt: see YouTube link
Movie Trailer: https://www.youtube.com/watch?v=8hwAjZlA4TY&t=6s

Content

In the apocalyptic film Independence Day, aliens attack earth. During the film, the viewer is told that the mother ship has a gigantic diameter of ¼ of the mass of the moon, and is located within a geostationary orbit (1/10 the distance between earth and moon, so 3.8×10^7 m) around the earth. What affect will the mother ship have on earth? In order to do this, calculate the ratio of the forces of gravity of moon and mother ship on earth.

Tasks

1) To what factor do the forces differ?

    Gravitational force F is determined using $F = G \frac{m_1 m_2}{r^2}$

    G is the gravitational constant with $f = 6.67 \times 10^{-11} \frac{m^3}{kg\ s^2}$

1) When looking at the effect that the gravitational force has on the tides, the above formula applies with r^3 instead of r^2. To what factor are the tides reinforced? How does this affect earth?

SOLUTION

1) Let m_1 be the mass of the moon, M the mass of earth and r the distance earth – moon.

   We are calculating the relation between the gravitational forces, so

   $$\frac{F_{Ship}}{F_{Moon}} = \frac{f\frac{0.25 \cdot m_1 M}{(0.1 \cdot r)^2}}{f\frac{m_1 M}{r^2}} = \frac{\frac{0.25 \cdot m_1 M}{0.1^2 \cdot r^2}}{\frac{m_1 M}{r^2}} = \frac{0.25 \cdot m_1 M \cdot r^2}{0.1^2 \cdot r^2 \cdot m_1 \cdot M} = 25$$

   This means, the effect on gravity is 25 times stronger from the space ship, compared to the moon.

2) Calculating the effect on high and low tide is somewhat more complicated and the details do not interest us at this point. What counts is that the radius takes affect with close to the third power:

   $$\frac{F_{Ship}}{F_{Moon}} = \frac{f\frac{0.25 \cdot m_1 M}{(0.1 \cdot r)^3}}{f\frac{m_1 M}{r^3}} = \frac{\frac{0.25 \cdot m_1 M}{0.1^3 \cdot r^3}}{\frac{m_1 M}{r^3}} = \frac{0.25 \cdot m_1 M \cdot r^3}{0.1^3 \cdot r^3 \cdot m_1 \cdot M} = 250$$

   The affect of high and low tide would therefore be 250 times greater. The space ship would not even have to attack, it could simple stay in its geostationary orbit for a few hours - everything else would take care of itself.

FLASH

Keywords: Kinetic Energy, Velocity, Braking Effort, Speed of Light,
Time of movie excerpt: see YouTube link
Movie Trailer: https://www.youtube.com/watch?v=82fanYF8i9E

Content

After lightning strike at a chemical laboratory, superhero Flash gains the power of running at the speed of light (300 000 km/s). The necessary energy for such an incredible feat obviously comes from a health breakfast. But how much does our hero have to eat to be this fast? As a matter of fact, in the comics Flash is actually not depicted as very muscly, but – since he is a runner – is drawn as rather slim and wiry. Let's assume he weighs about 70kg. As a measure we will use a hamburger, which contains about 273kcal or 1140kj.

Tasks

How many hamburgers does Flash have to eat, to get the required energy to comfortably run at 1% of the speed of light? $v_{Light} = 300.000 \frac{km}{s}$

SOLUTION

First of all, we calculate the kinetic energy to get to the given velocity of

$$0.01 \cdot 300000000 \, \frac{m}{s} = 3000000 \, \frac{m}{s}.$$

$$E_{kin} = \frac{1}{2}mv^2 = \frac{1}{2} \cdot 70 \text{ kg} \cdot \left(3000000 \, \frac{m}{s}\right)^2 = 3.15 \cdot 10^{14} \text{J}$$

One hamburger contains the energy of 1140 kilojoule = 1140000 Joule

Therefore, it applies: $\frac{3.15 \cdot 10^{14} \text{J}}{1140000 \text{J}} = 276315789$ hamburgers (so over 276 million).

Note: Hereby we are disregarding the relative increase in mass. It is also assumed that all and any energy contained in the hamburgers is transformed into kinetic energy.

SUPERMAN IV

Keywords: Rotation, Centrifugal Force, Angle Velocity, Stars
Time of movie excerpt: see Trailer
Movie Trailer: https://www.youtube.com/watch?v=jwuB2ub5aek

Content

In the film Superman IV – The Quest for Peace from 1987, Superman is dragged into the disarmament affairs. Triggered by a failed summit meeting and the worries of a little boy, the Kryptonian hero decides to collect all nuclear weapons on earth in a gigantic, 50m long net, and, like a hammer thrower, hurl it into the sun. Every rocket has a mass of about 2000kg. There are about 19 000 nuclear warheads on earth. In the movie excerpt, Superman needs about 4 seconds per rotation. About 14 seconds pass from the moment he lets go, until the point of impact on the sun.

Tasks

1) What centripetal force impacts Superman?

2) How long will it take for the net to hit the sun (distance: $1.5 \times 10^{11}\,m$), using the above time of rotation?

3) Calculate the velocity and angle velocity of the net, assuming, as shown in the film, it will take 14 seconds to hit the sun. Would this be possible?

SOLUTION

1) $F_c = \dfrac{m \cdot v^2}{r} = \dfrac{19.000 \cdot 2000 \, kg \cdot v^2}{50 \, m}$

The velocity is defined as $v = \dfrac{s}{t}$, since we are dealing with a circle it applies:

$$v = \frac{D}{t} = \frac{2 \cdot \pi \cdot r}{4s} = 78.54 \, \frac{m}{s}$$

$$F_c = 4.69 \times 10^9 \, N$$

2) $v = \dfrac{s}{t} \rightarrow t = \dfrac{s}{v} = \dfrac{1.5 \times 10^{11} m}{78.54 \frac{m}{s}} = 1,909,854,851 s = 60.56$ years.

3) $v = \dfrac{s}{t} = \dfrac{1.5 \times 10^{11} m}{14s} = 1.07 \times 10^{10} \, \dfrac{m}{s}$

This corresponds to 35 times the speed of light.

$\omega = \dfrac{v}{r} = \dfrac{1.07 \times 10^{10} m/s}{50m} = 214,000,000 \dfrac{1}{s}$ (degrees per second)

$T = \dfrac{2 \cdot \pi}{\omega} = 0.000000029 \, s$ (orbit period)

The Physics of Hollywood

STAR TREK

Keywords: Meeting point, Spaceship
Time of movie excerpt: see Trailer
Movie Trailer: https://www.youtube.com/watch?v=PiRN2Jz_2KI

Content

Star Trek was an American science fiction television series from 1960s. Under the command of Captain Kirk, the starship Enterprise explored unknown parts of the universe. A number of aspects of the show were controversial for viewers – the ship was operated by a Russian and an Asian. Furthermore, Star Trek was first to show a kiss between a black and a white person on television. The 6th cinema movie – alongside current events i.e. the end of the cold war – dealt with the topic of the nearing end of the Klingon Empire, and the fear of humanity faced with an unknown future.

Tasks

Assume the Enterprise takes 10 hours from earth to the Klingon home planet Oo'noS. A Klingon 'Warbird' takes 15 hours from Oo'noS to earth. After how many hours do the two ships meet if they both leave at the same time?

SOLUTION

First of all, we look at how far they get in one hour.

Enterprise: $\frac{1}{10}$ of the way; 1h

Warbird: $\frac{1}{15}$ of the way; 1h

After x hours, the two meet and have therefore – combined – covered the whole distance (=1), so:

$$\frac{x}{10} + \frac{x}{15} = 1$$

$$\frac{3x}{30} + \frac{2x}{30} = 1$$

$$\frac{5x}{30} = 1$$

$$5x = 30$$

$$x = 6$$

They meet after 6 hours.

AVENGERS

Keywords: Velocity, Trigonometry
Time of movie excerpt: 02:03:40 – 02:06:12
Movie Trailer: https://www.youtube.com/watch?v=eOrNdBpGMv8

Content

In the action movie Avengers, a group of superheroes fight against an alien invasion. During the finale, the US military fires a nuclear missile with the velocity v_1 towards the city centre, in order to stop the aliens. At the same time, superhero Ironman starts with the velocity v_2 to try and stop the missile.

Tasks

Draw a diagram and develop a formula for the dependency of the angle α to Iron Man's velocity.

SOLUTION

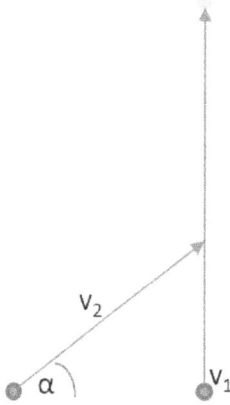

It applies: $s = v \cdot t$, let the distance of the starting points be d.
Therefore:
$s_1 = v_1 \cdot t$ and
$s_2 = v_2 \cdot t$ and
$Sin(\alpha) = \frac{s_1}{s_2} = \frac{v_1 \cdot t}{v_2 \cdot t} = \frac{v_1}{v_2}$.

The Physics of Hollywood

ARMAGEDDON

Keywords: Spaceship, Centrifugal Force, Gravity
Time of movie excerpt: 01:12:12 – 01:12:44
Movie Trailer: https://www.youtube.com/watch?v=kg_jH47u480

Content

During the course of action film Armageddon, the heroes are on board the (at this point already destroyed) Russian space station MIR. In order to save costs during shooting, the film forewent floating astronauts and, instead, claimed there was 'artificial gravity' on board of the station. This 'artificial gravity' can actually be achieved through the rotation of the space station. Hereby, the station rotates so quickly, that a human's weight is exactly cancelled out through the (outwardly working) centrifugal force. But since the basic unit of the MIR only had a diameter of 4.15m, which is very little when compared to the height of a human, the force would impact the cosmonauts' heads a lot less than their legs.

Tasks

1) Assuming we disregard these issues – how fast would the MIR have to rotate to generate artificial gravitation?
2) With this result, could a space shuttle dock?

SOLUTION

1. The centrifugal force has to be as large as gravity, therefore it applies:

$$F_G = F_c$$

$$m \cdot g = \frac{mv^2}{r}$$

$$g = \frac{v^2}{r}$$

$$v = \sqrt{g \cdot r}$$

$$v = 4.511 \frac{m}{s}$$

2. It is also interesting to look at how many rotations per second this responds to. The rotation period T hereby corresponds to the quotient of distance and velocity ($t = \frac{s}{v}$) with $s = 2\pi r$.

$$T = \frac{2\pi r}{4.511 \frac{m}{s}} = 2.89\ s$$

The station would have to rotate every three seconds – no construct would be able to withstand that, not to mention that it would be impossible for any space shuttle to dock on to the station.

TRANSFORMERS 2

Keywords: Centrifugal Force, Angle Velocity
Time of movie excerpt: 00:55:10 – 00:55:30
Movie Trailer: https://www.youtube.com/watch?v=dxQxgAfNzyE

Content

In the movie Transformers good alien robots (Transformers) are fighting evil alien robots (Decepticons), whereby a number of American cities are being wrecked. In Transformers 2: Revenge of the Fallen, the protagonists are trying to escape in a car, which is then lifted into the air by a helicopter. During transportation, the vehicle turns (T = 2s) around the axis of the rope, through this the passenger door opens. Sam (the hero) is thrown from the car and just manages to hold on to the door handle, at some point with just one hand.

Tasks

What force is impacting Sam during the rotation? Make appropriate assumptions. (Simplify and disregard gravity)
To compare: The current world record for weight lifting in Sam's weight class is 210kg, using both arms.

SOLUTION

It applies:

Centripetal force $F_c = m\omega^2 r$

and $\omega = \frac{2\pi}{T}$

Assumption:

$m = 80kg$ and $r = 1.5m$

It follows: $F_c = 80kg \cdot \left(\frac{2\pi}{T}\right)^2 \cdot r = 1,184.35N$

With both arms 2,200N are just possible. This means that it would be just about possible for a very strong man to hold on to the door, but not for someone who is not well trained.

The Physics of Hollywood

ARMAGEDDON

Keywords: Angle Velocity, Path Velocity, Centrifugal Force
Time of movie excerpt: see trailer
Movie Trailer: https://www.youtube.com/watch?v=OnoNITE-CLc

Content

In the action movie Armageddon two space shuttles start from a space port in Cape Canaveral to save the world.

There about a dozen space ports around the world – most of them are close to the equator. Why?

Tasks

1) Calculate the earth's angle velocity.

2) Calculate the earth's path velocity are the equator.

3) What is the path velocity at the poles?

4) Calculate the centripetal force impacting a rocket (m = 50t) at the equator. To how much N is it smaller at the equator compared to the north pole?

Helpful invariables:

- Earth's diameter: 6300km

- One day is 24h

SOLUTION

1) $\omega = \frac{2\pi}{T} = \frac{2\pi}{24 \cdot 60 \cdot 60} = 7.2 \times 10^{-5} s^{-1}$

2) $\omega = \frac{v}{r} \rightarrow v = \omega \cdot r = 7.2 \times 10^{-5} s^{-1} \cdot 6{,}300{,}000 m = 458 \frac{m}{s}$

3) Simplified, the path velocity at the poles is 0.

4) $F_c = \frac{mv^2}{r} = \frac{50{,}000 kg \cdot 458^2 \frac{m^2}{s^2}}{6{,}300{,}000 m} = 1{,}665\ N \rightarrow 169\ kg$

HULK

Keywords: Torque, Force, Pressure
Time of movie excerpt: 00:53:00 – 00:54:00
Movie Trailer: https://www.youtube.com/watch?v=xbqNb2PFKKA

Content

Through a gamma bomb of a communist spy, physicist Dr. Robert Bruce Banner has been turned in The Incredible Hulk, a two and half meter tall, green monster. In the cinema version from 2008, it can be seen how Hulk grabs an army Hummer (3.5 tons, 4.48 meters long), spins around and hurls it away. In order to lift such a car, Hulk would have to possess a certain weight himself. What is that weight?

Tasks

1) In order to do this, we need to determine the torque of the moving vehicle. The force's point of attack is at half the vehicle length.
   Since the car and Hulk rotate steadily, their torque values are the same.

2) Calculate Hulk's weight. The force's point of attack is at half of Hulk's diameter: 50cm

3) A standard ceiling can withstand pressure of up to 150kg/m^2. Calculate the pressure that The Incredible Hulk applies to the floor. Make appropriate assumptions. Would he be able to enter the house?

SOLUTION

1) Torque is the product of force and radius:

$$M = \vec{r} \times \vec{F} = 2.42m \times 35{,}000N = 84{,}700 \text{ Nm}$$

2) This time, we are looking for the weight force *F*, with the given torque value *M* and a second given radius of *0.5m*.

$$F = \frac{M}{r} = \frac{84{,}700\text{Nm}}{0.5\text{m}} = 169{,}400 \text{ N}$$

This corresponds to a weight of almost 17 tons.

3) Hulks foot has a surface of about $0.2m \cdot 0.4m = 0.08m^2$. Therefore the pressure is about $211{,}750 \frac{\text{kg}}{\text{m}^2}$.

No floor would be able to support him.

AIRWOLF

Keywords: Torque, Force, Pressure
Time of movie excerpt: 00:53:00 – 00:54:00
Movie Trailer: https://www.youtube.com/watch?v=FOFNOdKS8wE

Content

The television series Airwolf was shot in the 1980s and told the adventures of the secret combat helicopter and its pilot Stringfellow Hawke during the cold war. Hawke uses the miracle chopper to black mail the CIA: They are supposed to find free his brother, missing in action in the Vietnam war. While waiting for the organization to make good on this, he goes on secret missions and solves private problems – similarly to the Knight Rider Series about the speaking miracle car K.I.T.T.

Newton's third axiom states that every action generates a corresponding reaction. For a helicopter, the reaction to the rotor is the torque. The helicopter's hull therefore rotates in the opposite direction of the rotor. The tail rotor compensates this.

1) Why is the helicopter called 'helicopter'?

2) Look at the picture. Describe how the main and rear rotor work to keep the helicopter steady (How do the rear and main rotors work in steering the helicopter in a given direction?)

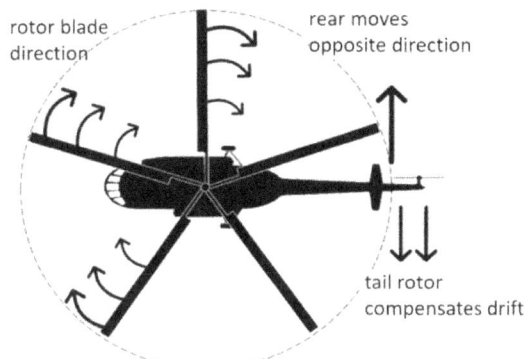

rotor blade direction

rear moves opposite direction

tail rotor compensates drift

3) Big cargo helicopters do not have a rear rotor, but two main rotors. How does this work? What is the advantage?

SOLUTION

1) The term 'helicopter' comes from the Greek words 'helix', meaning spiral, and 'pteron', meaning wing.

2) The main rotor generates a torque. If it rotates to the left, the helicopter underneath would turn to the right – the rear rotor has to cancel out this movement.

3) The two rotors have to move in opposite directions, in order to cancel out the torque. This way, the rear rotor is not needed anymore. The advantage of this is that there is double the lift and it is therefore possible to transport heavier loads.

HARRY POTTER

Keywords: Vacuum, Weight, Density
Time of movie excerpt: see Trailer
Movie Trailer: https://www.youtube.com/watch?v=x4hyAAV1kwQ

Content

In the immensely popular 'Harry Potter' series, the viewers accompany wizard student Harry Potter year after year during his time at Hogwarts School for Witchcraft and Wizardry. At the beginning of most of the films, Harry and his friends can be seen as they are searching for platform 9¾, while carrying their heavy suitcases.

Even though we do not possess magic abilities, natural sciences can still help us to make suitcases lighter.

'If the suitcase was filled with helium, it would have a slight lift and would be lighter' the chemist would say.

As physicists we go a step further: 'If the suitcase is 'filled' with vacuum, it would be even lighter, correct?'

Tasks

So which is it? A balloon filled with helium floats upwards. And what about an imaginary balloon filled with vacuum? Which would make Harry's suitcase lighter?

SOLUTION

Even helium has an atomic mass or weight. It might be lighter than that or 'air', but it is still a weight.
Therefore, it is heavier than 'nothing' – the suitcase should be filled with vacuum, only then it will be truly empty.

STAR TREK 1

Keywords: Spaceship, Angle velocity
Time of movie excerpt: see Trailer
Movie Trailer: https://www.youtube.com/watch?v=gxAaVqdz_Vk

Content

In the first Star Trek film from 1980, an alien space ship is threatening earth. During the film it becomes clear, that it is an old earth satellite from the 20th century.

Keeping satellites on course can be extremely difficult. In order to steer them, a flywheel is needed. When the wheel (see picture) is set rotating, the space ship turns in the opposite direction. When the wheel stops, the space ship stops turning, but stays at the new alignment under the angle $\Delta\theta_{new}$.

An insight into the complexity of this effect can be given when looking at the story of the space probe Voyager 2. The Voyager was launched in 1977 and is still sending data back to earth. However, while passing by the planet Uranus in 1986, it was put into unwanted rotation, every time a tape recorder, on board for data collection, was turned on. The engineers had to ignite the control nozzle every time this happened.

Assuming the tape recorder moved the probe during its 5-minute recording at an angle velocity of $6 \cdot 10^{-3}\mathrm{s}^{-1}$. At what angle does the probe drift off?

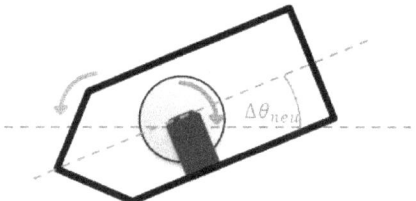

SOLUTION

It applies: $\varphi = \omega \cdot t$

Therefore: $\omega \cdot t = 6 \cdot 10^{-3} \frac{1}{s} \cdot 5 \cdot 60s = 1.8°$

The probe drifts off at an angle of $1.8°$

TED

Keywords: Pressure, Velocity, Free fall
Movie Trailer: https://www.youtube.com/watch?v=9fbo_pQvU7M

Content

The comedy film Ted (2012) starts off with a boy, wishing his teddy bear would come to life. This wish becomes reality and – soon grown-up – man ends up sharing a flat with his teddy. The movie's standards are clear from the poster, on which lead actor Mark Walhberg and his teddy are standing side by side in front of a row of urinals, the bear is holding a bear bottle in his hand.

As physicists we look at this poster from a nerd's perspective and asking ourselves the important question: How big does the distance d have to be in order to accurately hit the middle of the toilet?

Hereby we know: one urinates at an average pressure of 306 Pascal, at a height from about 1 meter. The toilet seat is about half a meter below and the density or urine is known to be $\rho_{Urine} = 1020 \frac{kg}{m^3}$.

Calculate the ideal distance d.

SOLUTION

The pressure with which one urinates, is hydrodynamic pressure. It applies:

$$p = \frac{1}{2}\rho v^2 \quad \rightarrow \quad v = \sqrt{\frac{2p}{\rho}} = \sqrt{2 \cdot \frac{306\,\text{Pa}}{1020\,\frac{\text{kg}}{\text{m}^3}}} = 0.77\,\frac{\text{m}}{\text{s}}$$

The height difference is 0.5 meters – a droplet of Urine therefore takes

$$s = \frac{1}{2}gt^2 \quad \rightarrow t = \sqrt{\frac{2s}{g}} = \sqrt{\frac{1\,\text{m}}{9.81\,\frac{\text{m}}{\text{s}^2}}} = 0.32\text{s}$$

In 0.32s the droplet travels exactly $s = v \cdot t = 0.77\,\frac{m}{s} \cdot 0.32\text{s} = 0.25\,\text{m}$

The ideal distance to the middle of the toilet bowl is 25cm.

MISSION IMPOSSIBLE 4

Keywords: Air pressure, Mass,
Time of movie excerpt: see YouTube link from 1:54 min.
Movie Trailer: https://www.youtube.com/watch?v=EDGYVFZxsXQ

Content

In the spy film Mission Impossible 4, Ethan Hunt has to save the world from an atomic war. During the film, he climbs up the outer wall of the Burj Khalifa, which is 828 meters high, using a special type of gloves. After one of the gloves fails to work, Ethan is left hanging on to the window pane for several seconds by just one hand.

Is this possible?

Tasks

From the Torricelli experiment we know, that air pressure can push a water column in the air to a maximum of 10.33m

1) With the aid of the Torricelli experiment, calculate the weight of the air column across a surface of $1m^2$.
2) Make appropriate assumptions. Is the surface of one glove large enough to hold Ethan Hunt?
3) Calculate the minimum surface needed to support his weight.

$$\rho_{Water} = 1000 \frac{kg}{m^3}$$

SOLUTION

1) The weight of the water column corresponds exactly to the weight of the air column. So: Ten kilos of air column can push up exactly ten kilos of water. A water column across 1 m² with a height of 10.33m, has a volume of 10.33m³.

   It applies:

   $$p = h \cdot g \cdot \rho$$
   $$1000\,\frac{kg}{m^3} \cdot 9.81\,\frac{m}{s^2} \cdot 10.33m = 100000\ Pa$$
   $$= 100000\ Nm^2$$
   $$= 10000\ kg/m^2$$

   The air column across 1m² therefore has a weight of about 10 tons. The hose in the experiment has a diameter of about 1cm². Accordingly, the air column above weighs 1kg.

2) Per cm² a weight of 1kg rests on the glove. A glove has a surface of about 90cm². Accordingly, the glove could support a weight of about 90kg. His stunt is possible.

3) In accordance with Ethan Hawk's weight, the surface has to be in cm².

ANT-MAN

Keywords: Reduction, Mass, potential energy
Time of movie excerpt: see Trailer
Movie Trailer: https://www.youtube.com/watch?v=QfOZWGLT1JM

Content

In the 2015 movie adaption of the comic book Ant-Man, former burglar Scott Lang is shrunk to ant size with the aid of a super suit.

Tasks

1) Ant man is shrunk to size of about 12mm, but still keeps the strength of a human. Calculate his mass at constant density. Make appropriate assumptions.

2) Considering the calculated mass and the strength of a normal human being, how high would Ant-Man be able to jump?

- First of all, devise a formula for the acceleration work W_B during a jump

- During a jump there are three stages: Squatting down (I) one accelerates until standing upright (II) and then jumps to a maximum height h (III). How does the potential energy change throughout the jump (compare stage (I) and (III))?

Equate both formulas. Calculate your own jumping power and use the result as the measure to answer the question of how high Ant-Man could jump.

SOLUTION

1) At an estimated height of 1.80m and weight of 80kg we will simplify Ant-Man's shape to a cylinder with a radius of 30cm and start with calculating his density:

$$\rho = \frac{m}{V} = \frac{80kg}{\pi \cdot r^2 \cdot h} = 157\frac{kg}{m^3}$$

Using this, as well as his shrunken height of 12mm and radius of 0.2 mm, we can calculate his new mass:

$$m = \rho \cdot V = 157\frac{kg}{m^3} \cdot \pi \cdot (0.0002m)^2 \cdot 0.012m = 2.36 \times 10^{-7}kg = 0.2mg$$

To compare: An ant has a mass of about 4mg

2) With the aid of the below diagram, we can see how the jumping process takes place:

SOLUTION

For the acceleration work it applies:

$$W_B = F_B \cdot s$$

The acceleration work is used to increase the potential energy of the jumping person:

This potential energy is calculated as follows:

$$E_{pot} = m \cdot g \cdot (h + s)$$

Through equating the two formulas one finds:

$$W_B = E_{pot}$$

$$F_B \cdot s = m \cdot g \cdot (h + s)$$

We are especially interested in the height of the jump h:

$$h = \frac{F_B \cdot s}{m \cdot g} - s \sim \frac{1{,}700 \, N \cdot 0.30m}{2.36 \times 10^{-7} kg \cdot 9.81 \frac{m}{s^2}} - 0.3m = 220{,}287{,}150m$$

To compare: This is a half of distance from earth to moon.

TERMINATOR GENISYS

Keywords: Free fall, Angle velocity
Time of movie excerpt: see trailer
Movie Trailer: https://www.youtube.com/watch?v=FqbOFjl7ZWE

Content

In the action movie Terminator Genisys from 2015, there is a spectacular chase between two helicopters. Arnold Schwarzenegger jumps out of the first helicopter about 40m down into the rotor of the second one in order to stop it. Neither do the rotor blades shatter, nor does the Terminator get sliced up by them, as visible in the following scenes. He simply falls through the blades.

Tasks

Is it possible for Arnold Schwarzenegger to fall through the running rotor, without getting hurt or damaging the blades? Make appropriate assumptions.

Hints: The helicopter's engine manages about 300 rotations/min and possesses a rotor circuit diameter of 15m.

SOLUTION

- First of all, we have to calculate how long the Terminator falls for:

$$s = \frac{1}{2}at^2 \ \rightarrow t = \sqrt{\frac{2s}{a}} = 2.86s$$

- Then we have to calculate the velocity he has by the time he collides with the helicopter:

$$v = a \cdot t = 28\frac{m}{s}$$

- Now we have to take a look at how long it takes for the Terminator to fall through the rotor. To simplify we assume that his body is 2m long and, during this short time span, displays a steady motion:

$$v = \frac{s}{t} \rightarrow t = \frac{s}{v} = \frac{2m}{28m/s} = 0.07s$$

  The terminator would need 0.07 seconds to fall through the rotor. The question now is how many times his body would be hit during this period.

- The helicopter does 300 rotations per minute. Since there are two blades we have to double this.

$$600\frac{U}{min} = 10\frac{U}{s}$$

  This means that every 0.1 second the rotor blade passes a given point.

It is therefore unlikely but technically possible for Schwarzenegger to fall through the blades.

The Physics of Hollywood

THE MARTIAN

Keywords: Radioactivity, Performance, Thermodynamics
Time of movie excerpt: 00:37:24 – 00:38:28
Movie Trailer: https://www.youtube.com/watch?v=ej3ioOneTy8

Content

The Martian (2015) tells the story of astronaut Mark Watney who is left on Mars following and accident, and now has to fight for survival.

To increase the range of his Mars rover, he turns off the heating. In order to not freeze to death on the cold planet, he digs up the radionuclide battery of an old satellite – loaded with 25kg of plutonium 238Pu – and uses it as a source of heat. Radionuclide batteries used in space travel obtain their energy from the radioactive decay of the plutonium, and can power a space probe for many decades. The thermal output resulting from this radioactive decay is 450 W/kg.

Tasks

1) Determine the thermal output of the battery at an efficiency of 8% and compare this to a conventional electric heater.

2) In the movie, Watney is sitting mere centimeters away from the battery. Research on the internet, whether this is realistic.

SOLUTION

1) The thermal output is $25 \cdot 450 \cdot 0.08 = 900$.
    900 Watts is comparable to a conventional electric heater.
2) Plutonium is an alpha emitter and easily shielded. The scene is realistic.

BATMAN VS SUPERMAN

Keywords: Relativistic energy, speed of light, energy
Time of movie excerpt: see YouTube link
Movie Trailer: https://www.youtube.com/watch?v=IwfUnkBfdZ4

Content

Batman vs Superman is the 2016 adaption of a comic book. Out of worry about the power of the uncontrollable, god-like Superman, Batman decides to try and destroy him. At the same time, the world is in discussion about what kind of superheroes earth ultimately needs.

During the course of the film, both superheroes start wildly bashing each other. But even Superman had to adhere to the rules of physics and cannot hit faster than at the speed of light c.

Tasks

1) Calculate the relativistic energy of Superman's fist (m = 300g), if he hits with v = 99% of c.
   In order to classify the result, one occasionally converts Joule into million tons of the explosive TNT (Mt). For this it applies: 1 J = $2,4 \cdot 10^{-16}$ Mt.

2) Compare and interpret Superman's punch looking at the destructive power of 'Little Boy', the atomic bomb that destroyed Hiroshima with about 13 kilotons of TNT-equivalent explosive force.

SOLUTION

1) To calculate the energy, we have to look at the theory of relativity, taking into account the Lorentz transformation: Since the fist is moving, we are not looking at a rest mass m_0.

   In the first step we have to calculate the distortion factor γ:

   $$\gamma = \frac{1}{\sqrt{1 - \frac{v^2}{c^2}}} = 7.09$$

   Using this, we can calculate the relativistic energy E:

   $E = m_0 \cdot \gamma \cdot c^2 = 1.9 \times 10^{17} J = 190\ 000\ 000\ 000\ 000\ 000$ Joule

2) $1.9 \times 10^{19} J \cong 45.6\ Mt = 45,600\ kt = 45,600,000$ tons TNT

   The impact of the punch corresponds to 3,500 times that of the Hiroshima bomb.

BATMAN VS SUPERMAN

Keywords: Relativistic energy, speed of light, energy
Time of movie excerpt: see YouTube link
Movie Trailer: https://www.youtube.com/watch?v=IwfUnkBfdZ4

Content

Batman vs Superman is the 2016 adaption of a comic book. During the course of the film, both superheroes start wildly bashing each other. In the above task you were able to calculate, that a single punch by Superman corresponds to 190 quadrillion Joule of energy.

Tasks

1) With the aid of Boltzman's constant, calculate the temperature generated by Superman's punch, under the assumption that all kinetic energy is converted into thermal energy.

2) Research on the internet. What temperature is there on the inside of the sun? At what degrees Kelvin do atoms disintegrate? What temperature might there have been during the big bang?
    Compare to 1) and make assumptions about the consequences of a punch by Superman.

SOLUTION

1) The average kinetic energy E_k of a particle is calculated using $E_k = \frac{3}{2} \cdot k \cdot T$ with Boltzmann's constant $k = 1.38 \cdot 10^{-23} \frac{J}{K}$. This results in a temperature of $9.2 \cdot 10^{39} \, °C$.

2) Temperature inside the sun: 15 000 000°C

   Temperature of nuclear fusion: 100 000 000°C

   Maximum possible temperature (Planck temperature): $1.4 \cdot 10^{32} \, °C$

   In a radius of a several hundred meters around Superman's punch, all atoms would evaporate completely.

PAY THE GHOST

Keywords: Earth rotation, Velocity
Time of movie excerpt: 01:14:06 – 01:15:30
Movie Trailer: https://www.youtube.com/watch?v=oVr5C4ysw3w

Content

In the supernatural thriller *Pay the Ghost*, Mike Lawford (Nicholas Cage) desperately tries to find his son, who went missing at a Halloween parade. In the thrilling conclusion of the film, Lawford follows his son through a gate and across a bridge into an Irish spirit world.

This gate is only open during Halloween and closes at the end of the day. The father grabs his son and flees from the change of date across the bridge, while it collapses behind him.

Tasks

Calculate: Is it possible to run away from the end of a day?
Hint: Earth has a radius of 6,370km.

SOLUTION

Earth has a radius of 6,370km, which corresponds to a circumference of

$C = 2 \cdot \pi \cdot r = 40,000$ km.

Since earth rotates once in a 24 hour period, the date line moves at

.

Nicholas Cage would never be able to run away from the change of date.

CAPTAIN AMERICA

Keywords: Free fall, Force, Acceleration
Movie Trailer: https://www.youtube.com/watch?v=1L3c17AmCZw

Content

In the superhero story Captain America: Civil War hero Steve Rogers fights his former companion-in-arms Bucky Barnes, who used to be Captain America's best friend during the second world war. After discovering Roger's secret identity as Captain America, Barnes is trained as his partner.

After a last mission towards the end of the war, he seems to have met his death. But a few decades later he resurfaces as brain washed soviet super agent Winter Soldier, and turns out to be the enemy of Captain America and the Avengers. His name 'Winter Soldier' comes from the fact that he is usually kept frozen in cryogenic stasis, and only activated to attack, and has thus effectively only aged a few years over several decades.

In the film, Barnes jumps off a highway bridge, about 9m high, landing on the street below completely unharmed. Is this possible?

1) Determine the falling time and use it to calculate the resulting velocity v_1.

2) Bucky only bends his knees a little bit after his jump, and needs 0.125s to absorb the impact. Calculate the negative velocity impacting his body.

3) Compare your result with the graph below, which depicts the limits of survivable forces. Is his jump realistic?

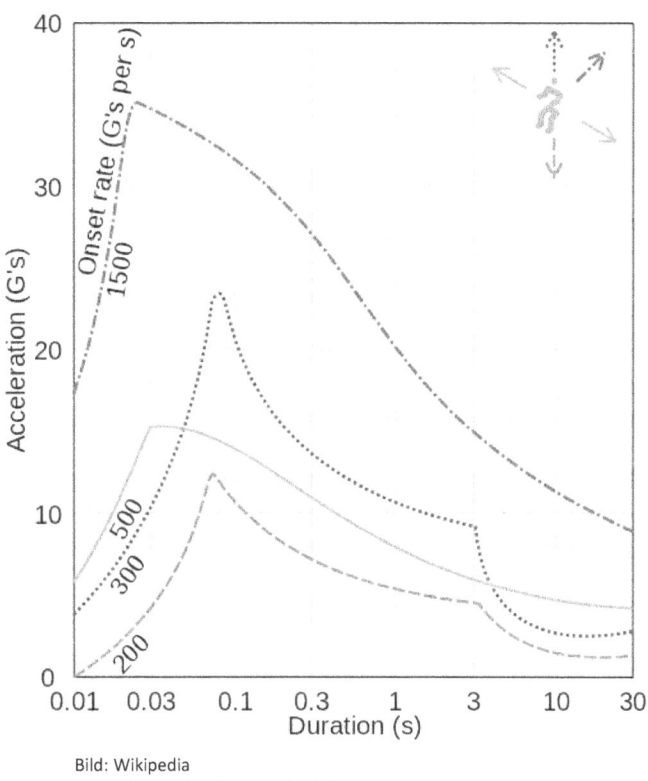

Bild: Wikipedia
https://en.wikipedia.org/wiki/G-force

SOLUTION

1) The falling time can be estimated by paying attention to the animation and is about

   t = 1.35s

   Therefore, it applies: $s = \frac{1}{2} \cdot a \cdot t^2$ and $a = v \cdot t$ from which follows $v = 13.24 \frac{m}{s}$.

2) Now we are looking for the acceleration a. We assume that his braking distance is 0.3m:

$$a = \frac{2s}{t^2} = \frac{0.6m}{(0.125s)^2} = 38.4 \frac{m}{s^2} = 4G$$

3) It is very possible to survive this jump, which is illustrated by the French sport *parkour*. However, imitation is strictly discouraged!

GREY'S ANATOMY

Keywords: Body, Free fall,
Time of movie excerpt: Greys Anatomy, 5x01, 39:30 – 40:46 (or YouTube link)
Movie Trailer: https://www.youtube.com/watch?v=FyePUT2pbIY

Content

Grey's Anatomy is an American hospital series, which is comparable to cheap TV soaps in terms of the frequency of tragedies each character has to endure on a regular basis. In the first episode of series 5, Dr. Yang falls on slippery ground and is impaled by an icicle.

Tasks

From a physical point of view, we have to question whether this scene is realistic. It is clear that the icicle falls onto the doctor from a height of about 4m, in the shape of a blunt cone 40cm in length and with a diameter of about 10cm. The human abdomen can withstand a pressure of $9 \cdot 10^6$ Pascal before it is pierced.
Complement the missing assumptions and decide: Is this realistic?

SOLUTION

First of all we have to determine the cone's volume: $V = \frac{\pi}{3} \cdot r^2 \cdot h = 1.05 \, dm^3$

This is comparable to 1kg in weight.

The abdominal wall can yield about 5cm (breaking distance), through which we can equate the icicles potential energy with the required breaking effort, in order to determine the necessary force:

$$m \cdot g \cdot h = F \cdot s$$

From this immediately follows:

$$F = \frac{m \cdot g \cdot h}{s} = 824 \, N$$

The human abdomen can withstand a pressure of $9 \cdot 10^6$ Pascal, which, when converted, corresponds to 900 Newton per square centimeter.

As long as the point of the icicle is not smaller than $1cm^2$, it most likely would not be able to penetrate the abdominal wall. However, the scene is not entirely unrealistic.

TRUE LIES

Keywords: Free fall, Acceleration, Air resistance, Pressure, Breaking effort
Time of movie excerpt: YouTube link
Movie Trailer: https://www.youtube.com/watch?v=i1CPsVNUAG8

Content

True Lies is a spy comedy with Arnold Schwarzenegger. The film is about a secret agent, who has to keep his job a secret from his family. His wife is bored by his (pretend) conventionality and starts and affair with a philanderer who tells her he is a secret agent. During the course of the film, the family get involved in a real altercation with an Arabic terrorist. Towards the end of the film, the bad guys celebrate their plan and randomly fire their machine guns into the air.

Tasks

The scene beckons the question of what happens to the projectiles after they have been fired into the air.

1) A projectile with a caliber of 7.62mm leaves a machine gun with a velocity of up to $800 \frac{m}{s}$. Calculate the possible height and decide: Is the result realistic?

2) Experimentally, a velocity of 500kph has been proven for a falling projectile. Determine the kinetic energy of such a projectile.

3) Determine the danger coming from a falling projectile. According to research, the top of the skull can withstand a pressure of 39.1 MPa.

SOLUTION

1) Let the throwing height be: $y_{max} = \frac{v_0^2}{2 \cdot g} = 31{,}408 \, m \approx 31 \, km$.

No, the result is not realistic. At this point we need to take the air resistance into consideration. In reality such a projectile reaches a height of 2,500m and falls back down at a velocity of 500kph.

For the kinetic energy it applies:

$$E_k = \frac{1}{2} \cdot m \cdot v^2 = \frac{1}{2} \cdot 0.012 \, kg \cdot \left(139 \frac{m}{s}\right)^2 = 115 \, J$$

2) We assume that the top of the skull can yield about 1cm before it breaks – meaning the breaking distance is 0.01m.

$$E_k = F \cdot s$$

$$F = \frac{E_k}{s} = \frac{115 \, J}{0.01 \, m} = 11{,}500 \, N$$

The skullcap can withstand a pressure of about 39.1MPa, which corresponds to 39,100,000 Newton per square meter, or 3,910N per square centimeter. The projectile hits the head with a force of about 11,500 Newton – since the tip is likely smaller than 3cm^2, it would penetrate the skull and lead to serious, most likely fatal injuries.

Experimentally, the danger of such shots into the air have been proven.

Further question: Why is it allowed for hunters to shoot flying ducks out of the sky?

THOR 3

Keywords: Momentum
Time of movie excerpt: see trailer
Movie Trailer: https://www.youtube.com/watch?v=F7ayGFHGqeQ

Content

In the upcoming third film about superhero Thor, the trailer already depicts a fight between Thor and the Incredible Hulk. In the last second it is visible how the two are running towards each other, while Thor is already raising his fist for a punch.

Tasks

Watch the trailer and explain: What will the next few seconds after the collision look like?

SOLUTION

According to the pictures, the incredible Hulk as a way larger mass than Thor. According to conservation of momentum it applies:

$$m_1 \cdot v_1 = m_2 \cdot v_2$$

Consequently, Thor would be hurled backwards with a considerable velocity, while Hulk would, at worst, remain in the same position.

WONDER WOMAN

Keywords: Impulse, Speed
Time of movie excerpt: s. Trailer
Movie Trailer: https://www.youtube.com/watch?v=1Q8fG0TtVAY

Content

With Wonder Woman, the first female superhero finally made her cinema appearance in 2017. As Amazon princess Diana she fights for justice in the First World War. In the trailer you can see how Wonder Woman fends off the rifle shots of German soldiers with her silver arm splints. In view of the many film scenes in which people hit by bullets are thrown backwards, we want to take a closer look at this sequence.

Tasks

1) Calculate the impulse that acts on Wonder Woman when a bullet weighs 5 grams and is 960 m/s fast and flies back in the same direction as in the trailer.

2) At what speed does Wonder Woman (frictionless) move backwards when pushed by the bullet? Make appropriate assumptions.

3) Does this also work with a machine gun? Assume that such a weapon from the 1st World War can fire about 500 rounds per minute.

SOLUTION

1) Calculate the impulse that acts on Wonder Woman when a bullet weighs 5 grams and is 960 m/s fast and flies back in the same direction as in the trailer.

$$\Delta p_{\text{bullet}} = p_{there} - p_{\text{back}} = m_K \cdot v_K - m_K \cdot (-v_k) = 2 \cdot m_K \cdot v_K$$
$$\Delta p_{\text{bullet}} = 2 \cdot 0.005\ kg \cdot 960\,\frac{m}{s} = 9.6\,\frac{m \cdot kg}{s}$$

2) At what speed does Wonder Woman (frictionless) move backwards when pushed by the bullet? Make appropriate assumptions.

Assumption: $m_{WonderWoman} = 50\ kg$

$$p = m \cdot v \quad \leftrightarrow \quad v = \frac{p}{m} = \frac{9.6\,\frac{m \cdot kg}{s}}{50\ kg} = 0.192\,\frac{m}{s} = 0.7\,\frac{km}{h}$$

Wonder Woman can easily withstand a bullet shot.

3) Does that work on a machine gun, too? Assume that such a World War I weapon can fire about 500 rounds per minute.

$$F = \frac{\Delta p}{\Delta t} = \frac{9.6\,\frac{m \cdot kg}{s}}{0.12\ s} = 80\ N$$

That's the weight of a toddler sitting on you. Wonder Woman shouldn't have any problems with that.

IT

Keywords: Performance, Velocity
Tiempoextrat: s. Trailer
Movie Trailer: https://www.youtube.com/watch?v=xKJmEC5ieOk

Content

In the horror movie IT (2017), the place Derry is haunted by a monster in the guise of the clown Pennywise. In the course of the film you can see the clown crawling out of a slide projection into reality in huge stature, he reaches - crawling on all fours - up to the ceiling of the garage. For an adult, the thigh bone will be able to withstand about 8,000 N, a vertebra (which only has to carry half of the body weight) about 3,600N.

Tasks

a) Make suitable assumptions. Are the bones able to support Pennywise after it is magnified?

b) Which size could she be at maximum before its bones would break?

c) The pitch (frequency) of the human voice is about 200 vibrations per second. To what factor would Penny wises vibration period change? (Hint: Look at the vocal chords as a thread pendulum)

d) The human hearing ability ranges from 20 to 20 000 vibrations. Would we still be able to understand Pennywise?

e) Simplified, hearing ability depends on the size of the ear drum (Hint: Surface!). So in reverse, would Pennywise be able to still understand normal sized humans?

SOLUTION

1) First of all, one has to establish the magnification factor x.
$$1.80 \cdot x = 7 \quad \rightarrow x = 3.9$$

This factor applies to length, width and depth. Penny wise's weight (volume), therefore changes with the third power:
$$80\text{kg} \cdot 3.9^3 = 4{,}705 \text{ kg.}$$

The bones' cross section (surface), however, only growth with the second power. The weight-bearing capacity is therefore:
$$8{,}000\text{N} \cdot 3.9^2 = 115{,}520\text{N} \approx 12\text{t}$$
$$3{,}600\text{N} \cdot 3.9^2 = 54{,}756\text{N} \approx 5.6 \text{ t}$$

Both bones are therefore able to support Penny wise's weight (note: The vertebrae only has to carry half the weight)

2) Wanted is the maximum magnification factor x.

It applies: $800\text{N} \cdot x^3 = 8{,}000N \cdot x^2$ or $\quad 400\text{N} \cdot x^3 = 3{,}600N \cdot x^2$

(The growing weight force can be no more than the maximum weight-bearing capacity) It follows that $x_O = 10$ and $x_R = 9$

Therefore, Pennywise could be $1.80\text{m} \cdot 9 = 16.2$ m tall at maximum.

3) For the vibration period of a pendulum it applies: $T = 2\pi \sqrt{\frac{l}{g}}$. Thus, the only

changeable variable is l. For the change in vibration period it applies: $T_{tall} = 2\pi \sqrt{\frac{3l}{g}}$.

The change factor is therefore $\sqrt{3} \sim 1.7$. The vibration period T is this smaller by a factor of 1,7. (Note: When viewing the vocal cords as surfaces rather than lines, the factor would have to be squared!)

4) It applies: $T = \frac{1}{f}$. The frequency is this larger by a factor of 1.7.
$200Hz : 1.7 = 117\ Hz$. Yes, we could still understand it.

5) Simplified, the ear drum can be viewed as a surface – therefore, it shrinks at a factor of 1.7^2. Accordingly, the audible frequency increases by a factor of 3^2. The audible range would lie between 2Hz and 2222Hz.

INTERSTELLAR

Keywords: Time Dilation, Black Hole, Relativity.
Time of movie excerpt: s. Trailer
Movie Trailer: https://www.youtube.com/watch?v=lznM-fygfqo

Content

Interstellar is considered one of the films with the best scientific content. In this participated the renowned theoretical physicist Kip Thorne who collaborated in the original version of the script. Cooper (Matthew McConaughey), a former pilot and engineer where the apocalyptic circumstances of the Earth force him to be a farmer, is part of the space exploration that is emerging as the only salvation of humanity. At a critical moment, they reach the planet Miller that is very close to Gargantua (black hole); here they must not only face the massive dangers of gravity, but also the relativistic effects on time.

Tasks

1) Before arriving at the planet Miller, astronaut Romilly mentions that 1 hour on the planet equals 7 years on the ship. How far, from the black hole, should the planet be found to notice this relativistic effect?

2) Is it physically possible for a planet to be at the distance given in 1)?

3) What does this dilation of time tell us about the black hole?

SOLUTION

1) The equation that relates time dilation by gravitational effects is

$$t_0 = t_f\sqrt{1 - \frac{2GM}{rc^2}} \to r = \frac{2GM}{\left[1-\left(\frac{t_0}{t_f}\right)^2\right]c^2} = \frac{2*\left(6.67*10^{-11}\frac{Nm^2}{kg^2}\right)*\left(1.989*10^{38}kg\right)}{\left[1-\left(\frac{24\,h}{61,320\,h}\right)^2\right]\left(3*10^8\frac{m}{s}\right)^2} = 8.84 * 10^8 km$$

Note: The mass of Gargantua is about 100 million times the mass of the sun.

2) To know if the distance found in 1) is physically possible, it is necessary to calculate the Schwarzschild radius

$$r = \frac{2GM}{c^2} = \frac{2*\left(6.67*10^{-11}\frac{Nm^2}{kg^2}\right)*\left(1.989*10^{38}kg\right)}{\left(3*10^8\frac{m}{s}\right)^2} = 2.94 * 10^8 km$$

3) Any object that is less than the Schwarzschild radius will be part of the black hole, therefore, the answer given in 1) is physically possible.

X-MEN

Keywords: Magnetic Field, Magnetic Force, Levitation
Time of movie excerpt: see YouTube link
Movie Trailer: https://www.youtube.com/watch?v=lznM-fygfqo

Content

For the year of 1963 the scriptwriter and American publisher, Stan Lee, came up with the brilliant idea of using the genetic mutation for the creation of beings with superpowers: the X-Men. In that same year appears Magneto, character par excellence of the strip well known for being the master of magnetism. Its mutant power consists, mainly, in generating and manipulating magnetic fields. His ability to manipulate and create magnetism not only allows him to completely raise a baseball stadium or the Golden State, it also allows him to levitate through the air.

Tasks

1) Is it physically possible for a person to levitate by the action of a magnetic field as it happens with Magneto? Justify

2) What is the intensity of the magnetic field that allows Magneto to move through the air?

3) How intense should the field generated by Magneto be that allows his to raise a plane?

4) What physical repercussions could be caused by exposure to such intense magnetic fields?

1) Diamagnetics are those materials that respond weakly to magnetic fields and always opposing them. The human being can be considered diamagnetic, due to its majority composition of water (75%) and carbon (19.37%). Thus, under the presence of a strong magnetic field, the human body would begin to feel the magnetic effects.

2) When applying a magnetic field $\vec{B}$, the substances produce a magnetization $\vec{M}$ given by:

$$\vec{M} = \frac{\chi V}{\mu_o}\vec{B},$$

χ is the magnetic susceptibility of the material (-9.5×10^{-6} for the human body), μ_o is the magnetic permeability in vacuum ($4\pi \times 10^{-7} N/A^2$) and V is the volume for the body. In the one-dimensional case, the magnetic moment is a dipole and the force felt by the external field is given by

$$f_{magnetic} = \frac{\chi V}{2\mu_o}\frac{dB^2}{dz}$$

Likewise, if we assume a smooth gradient field, that is, a uniform magnetic field, we can approximate the derivative as $\frac{dB^2}{dz} \sim -\frac{B^2}{L}$ (linear approximation by Taylor series). That which allows to rewrite the expression for magnetic force as

$$f_{magnética} = \frac{\chi V}{2\mu_o}\frac{B^2}{L}$$

In order for Magneto to levitate, it is necessary to comply $f_{Magnetic} > f_{gravitational}$. ($f_{gravitational} = mg = \rho V g$)

$$\frac{\chi V}{2\mu_o}\frac{B^2}{L} > \rho V g,$$

Solving for B

$$B^2 > \frac{2\mu_o \rho g}{\chi}L$$

We get that $g = 9.81 \, m/s^2$, $\rho = 980\frac{kg}{m^3}$ and $L = 1.88m$ (Magneto height), the magnetic field should be:

$$B > 69.14 \, T$$

SOLUTION

3) Let's analyze the case of a commercial aircraft, the Boeing 777. Its composition is 70% aluminum, 11% steel, 11% CFRP compounds(carbon-fiber-reinforced polymer), 7% titanium and 1% misc

$$\chi_{airplane} \approx 0.7\chi_{Al} + 0.11\chi_{Fe} + 0.11\chi_{CFRP} + 0.07\chi_{Ti}$$

$(\chi_{Al} = 2.07 \times 10^{-5}, \chi_{Fe} = 0.002, \chi_{CFRP} = -5.3 \times 10^{-7} \text{ y } \chi_{Ti} = 1.90 \times 10^{-3})$.

$$\chi_{airplane} \approx 6.97 \times 10^{-4}.$$

Similarly, we can estimate the average density

$$\rho_{airplane} \approx 0.7\rho_{Al} + 0.11\rho_{Fe} + 0.11\rho_{CFRP} + 0.07\rho_{Ti}$$
$$\approx 0.7(2.7 \ g/cm^3) + 0.11(7.9 \ g/cm^3) + 0.11(1.6 \ g/cm^3) + 0.07(4.5 \ g/cm^3)$$
$$\approx 3.25 \ g/cm^3 = 3250 \ Kg/m^3$$

Using again the expression for the magnetic field obtained in 2)

$$B^2 > \frac{2(4\pi \times 10^{-7} N/A^2)(3250 \ Kg/m^3)(9.81 \ m/s^2)}{6.97 \times 10^{-4}} L \approx (114.96 T/m)\text{L}$$

The total height of the plane is $18.6 \ m$, so if you want to lift it while the plane is horizontal, it's magnetic field would be

$$B > 46.24T$$

On the other hand, the approximate length of the plane is $63.7 \ m$. If the plane is in an upright position, Magneto would have to generate a magnetic field of

$$B > 85.58T$$

Copyright Notice

All films and characters mentioned in this book are protected by copyright and used to illustrate in a scientific and educational setting.

Thanks

Thank you so much for purchasing this book!
We hope it brought you some joy and triggered some inspirational ideas for your lessons.

If you find a minute to spare, We would greatly appreciate a review on Amazon. This book was developed without the help from publishers and without your reviews it will likely remain at around four sold copies. We are just teachers.

We are always happy to receive notes, corrections and ideas, so feel free to contact us on halbtagsblog@gmail.com and scardena.pr@gmail.com. (especially with regard to the translation.)

www.ingramcontent.com/pod-product-compliance
Lightning Source LLC
Chambersburg PA
CBHW081514220526
45467CB00010B/2917